Human Genetics

A Series
of Books
in Biology

CEDRIC I. DAVERN
Editor

Human Genetics

An Introduction to the Principles of Heredity

SAM SINGER

University of California, Santa Cruz

W. H. Freeman and Company
San Francisco

Cover:
Normal and sickled red blood cells from a person who has sickle-cell trait. Electron micrograph courtesy of Patricia N. Farnsworth.

Library of Congress Cataloging in Publication Data

Singer, Sam, 1944–
Human genetics.

(A series of books in biology)
Includes index.
1. Human genetics. I. Title.
QH431.S638 573.2′1 78–82
ISBN 0–7167–0054–9

Printed in the United States of Ameria

5 6 7 8 9 10 KP 0 8 9 8 7 6 5 4 3 2

Contents

Preface

Most people are naturally interested in the subject of genetics, especially as it relates to the inheritance of physical and behavioral traits that run in human families. Nonetheless, genetics has the reputation of being a difficult subject, and many people deprive themselves of the fun of understanding how genetics works because they wrongly assume that the subject is too complicated for them. I have written this book in the hope of sharing with you what I think are the fundamentals, and some of the interesting highlights, of human genetics. Anyone with enough interest in the subject to pick up this book and browse through it will probably be able to understand what is written here, regardless of previous background in biological science.

The book has five chapters. The first two are concerned with how the inheritance of some characteristics that run in human family lines can be explained in cellular terms, and with how geneticists explain the fact that the human population is made up of nearly equal numbers of men and women. Chapter 3 explains how the hereditary material, DNA, brings about its effects, and how some human characteristics can be explained as the results of chemical changes in DNA molecules. Chapter 4 is about the genetics of human populations. Here the emphasis is on the concept of "biological race" and on how heritable changes, or *mutations,* ultimately arise because of accidents that affect DNA molecules. Finally, chapter 5 explains how human characteristics, including those that pertain to behavior, oftentimes depend on the interactions of genetic and environmental factors, and then discusses some of the ways in which people can or could directly influence the genetic future of the human species.

For those who enjoy such things, the text is followed by some problems that pertain to the patterns of inheritance discussed in Chapters one and two. Answers and explanations are also provided. Anyone wanting to read more about human biology may be interested in a book entitled *The Biology of People* that I have written in collaboration with Henry R. Hilgard. (I have, in fact, taken the present volume from the larger work.)

I should sincerely like to thank the following professors who read over, commented on, or otherwise helped to improve my rough drafts: Henry R. Hilgard, Cedric I. Davern, Robert S. Edgar, and Ursula W. Goodenough. Thanks are also due to my friends at W. H. Freeman and Company, especially to Linda Chaput, Gunder Hefta, and John Painter. And finally, thanks to everyone who reads this book and thereby learns something, as I surely did while writing it.

Felton, California
January 1978

SAM SINGER

Human
Genetics

The members of the Augustinian monastery in old Brno, Czechoslovakia in the early 1860s. Gregor Mendel is third from the right. (Photo courtesy of Dr. V. Orel of the Moravian Museum, Brno.)

CHAPTER 1

Traits and Chromosomes

In the mid-eighteenth century the city of Paris was the scene of a most unusual mating that aroused widespread public interest. The affair concerned a male rabbit who unexplainably showed great sexual interest in a certain barnyard hen. The hen, for her part in the matter, readily tolerated the rabbit's advances but would have nothing to do with roosters. These two unusual animals, both of whom belonged to a disconcerted clergyman, were observed to "mate" frequently, but the naturalists of the day doubted whether the union of the two was as complete as that of a rooster with a hen or a rabbit with another rabbit. So when the hen obligingly laid six normal-looking eggs, there was great excitement. What would hatch out?

Some people expected long-eared furry chickens to result; others, rabbits with beaks and feathers. But to the great disappointment of most neither rabbit nor chicken nor anything in between emerged. The well-watched eggs merely sat and decomposed.

The eggs failed to develop because rabbits and chickens are different species of animals. As you may know, *species* may be defined as populations of organisms that retain their individuality in nature because they are reproductively isolated from other species around them. In general, reproductive isolation among animal species has two important and interrelated aspects: behavior and genetics.

2

Traits and Chromosomes

The Parisian observers of the ill-fated "mating" of the rabbit and the hen were as familiar with the behavioral aspects of reproductive isolation as we are. They were well aware that animals of different species generally show no interest whatsoever in mating with one another. But what the Parisians did not realize was that even if the behavioral aspect of reproductive isolation occasionally goes awry, the individuality of a species is still protected because species are genetically distinct from one another. What exactly does this mean?

Animals produce sex cells of two different types—eggs from the female and sperm from the male. The genetic uniqueness of sexually-reproducing species (and this includes virtually all animals) may be thought of as having its basis in different blueprints, or programs, for the elaboration of different species from fertilized eggs. As we shall see, each egg and each sperm usually contain within their nuclei half of the information necessary to set into motion the complicated process of elaborating a particular kind of animal according to the program of the species to which the parents belong. But in order for proper development to occur, both sex cells that merge to form the fertilized egg, or *zygote*, must contain the same basic program.

When the Parisian rabbit and hen mated it is not likely that their union resulted in a fertilized egg. This is because reproductive isolation operates even in sex cells, and rabbit sperm would be unlikely to penetrate and fertilize chicken eggs. In fact, even if we forced a rabbit sperm to fertilize a chicken egg by accurate mechanical injection of the sperm, the mating still would not produc feathered rabbits or long-eared chickens. A chicken egg fertilized in this way ha received two conflicting programs, one for constructing a chicken and one for constructing a rabbit. And the programs for each are different enough that the artificially fertilized egg burns up its supply of intracellular fuel and then dies in the confusion of attempting to initiate the development of a composite creature from conflicting plans.

Actually, rare instances of successful interspecies mating do occur, not only among animals in experimental circumstances, but in nature too. (Also, interspecies crosses are much more common among plants than among animals.) Generally, such crosses occur only between species that are closely related by evolutionary descent, and that therefore presumably have similar genetic

1–1
The mule, left, is a familiar hybrid that is produced by the mating of a female horse with a male donkey. The hinny, right, is more horselike in appearance and results from the mating of a female donkey with a male horse. (From "The Mule," by Theodore H. Savory. Copyright © 1970 by Scientific American, Inc. All rights reserved.)

programs. Perhaps the most familiar animal issuing from an interspecies cross is the mule, the offspring of the mating of a female horse with a male donkey, but viable crosses also occur among closely related species of fish, birds, and porpoises, among others (Figure 1-1). But the animals resulting from these interspecific crosses are often incapable of reproduction themselves and they are clearly exceptions to the rule of reproductive isolation.

Differences in genetic programs between species are in large part responsible for the fact that animal species are usually morphologically distinct: that is, one can usually tell different species apart merely by looking at them. But then it is possible in many instances to distinguish animals from one another *within* a species, too (intraspecific variation). This is especially true of land-dwelling vertebrates, and is nowhere more obvious than in the human species, which is by far the most variable species known. Human beings have various skin colors ranging from almost pure white to jet black, have head and body hair ranging from perfectly straight to tightly kinked, and have unique fingerprints and faces—except for identical twins, who, as we shall see, have identical genetic programs. Yet within the human species, as is true of all others, the characteristics of individuals are not randomly distributed throughout the entire population. There are clear-cut geographic and racial differences as well as differences between related family lines.

The science of genetics is concerned with the study of heritable differences, both how they originate and how they relate to an individual's genetic program. Genetics also concerns itself with the biological basis of the transmission of traits in lineages and with the distribution of heritable characteristics within the populations that make up a given species. In chapters to come we will investigate the biochemical basis of heritable traits in individual people and see how at least some of these traits may have come to have the distributions we observe within the human population today. But in this chapter our main concern is with identifying and explaining the simple patterns of inheritance shown by some rather clear-cut human characteristics that obviously run in families. For the most part, human patterns of inheritance can best be described by some basically simple—yet decidedly unobvious—principles first worked out in the 1860s by an Augustinian monk named Gregor Mendel (frontispiece).

What Mendel Did

Mendel discovered the basic patterns of inheritance by performing carefully planned experiments on the common garden pea, and his success was partly due to his wise choice of experimental subject. Pea plants are good subjects for simple genetic experiments for several reasons. First of all, individual pea plants have clear-cut differences in some easily recognizable alternate characteristics. For example, the ripe seeds may be either smooth or wrinkled, and either yellow or an intense green. Mendel chose to experiment with seven clearly alternate traits in his search for patterns in the way such traits are passed on from parents to offspring (Figure 1-2).

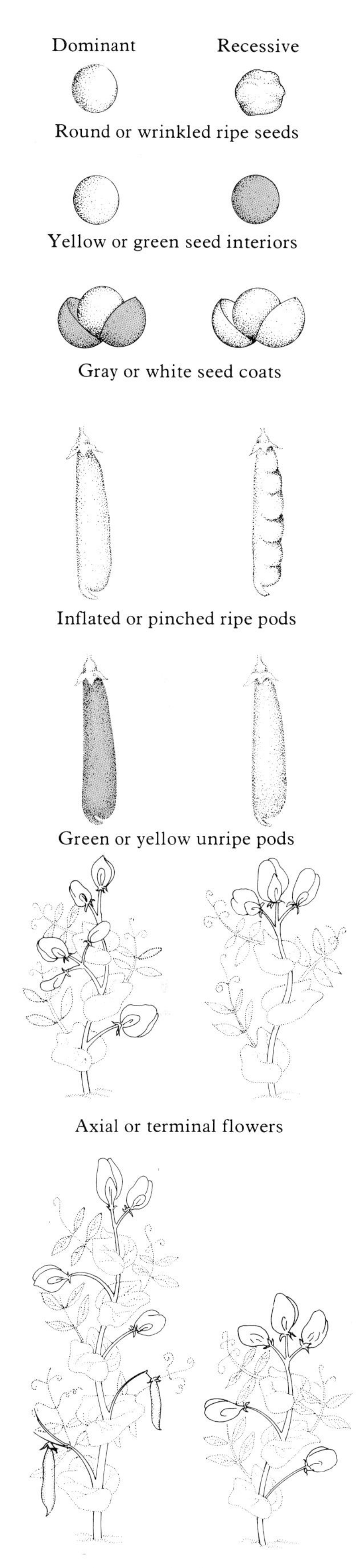

1–2
Mendel's early experiments were conducted upon seven pairs of alternate characteristics of garden pea plants.

Another reason pea plants make good experimental subjects is that they are self-fertilizing. That is, pea blossoms are so constructed that the male sex cells, which are contained in pollen grains, and the female sex cells, or eggs, are located in the same blossoms. (In other species, male and female sex cells may be produced separately, either by separate male and female flowers on the same plant or by flowers on separate male and female plants.) Self-fertilizing plants tend to breed "true," which means that their offspring usually resemble the parents exactly, at least in the alternate traits Mendel observed and recorded. True-breeding plants are thus good subjects for crosses of individual plants that differ in one or more alternate traits.

Mendel crossed plants that bred true for alternate traits and carefully recorded the distribution of these traits in their offspring. What he found is best illustrated by a recounting of some of his experiments.

Mendel began with two varieties of true-breeding pea plants, whose self-fertilized (self-pollinated) offspring had ripe seeds that were either round or wrinkled. He pinched off the pollen-producing parts (anthers) of each blossom on plants that produced only wrinkled seeds, and then fertilized the blossoms with pollen from plants that bred true for round seeds. (He also fertilized some "round blossoms" with "wrinkled pollen" and produced essentially the same results discussed in the following sentences.) Mendel then tied little paper bags over the blossoms to prevent any wind-borne or insect-borne pollen from contacting the artificially fertilized plants. When he opened the pods of his experimental plants he found that all of the seeds were round. The alternate trait, "wrinkled," seemed to have disappeared in the first generation of progeny produced from the cross, the F_1 generation. Mendel then planted the round seeds produced in the cross and allowed the resulting plants to fertilize themselves, as they usually do. When he examined the seeds produced by the second generation (the F_2 generation), he sometimes found round and wrinkled seeds lying together in the same pod (Figure 1-3). To be more exact, he found that about 25 percent of the total number of seeds were wrinkled. The trait "wrinkled," which had disappeared in the F_1 generation, had once again turned up in the F_2 generation about 25 percent of the time.

As shown in Table 1-1, Mendel found the same pattern for all seven of the traits he studied. For example, when true-breeding plants that had yellow seeds were crossed with those whose seeds were green, only yellow-seeded offspring were produced. Accordingly, Mendel called the member of the alternate pair of characteristics that showed up in all of the offspring of the F_1 generation, and in about 75 percent of the offspring in the F_2 generation, a *dominant trait.* And he

TABLE 1–1

Mendel's results from crosses involving some alternate characteristics of the common garden pea.

PARENT CHARACTERISTICS	F_1	F_2	F_2 RATIO
1. Round × wrinkled seeds	All round	5,474 round : 1,850 wrinkled	2.96 : 1
2. Yellow × green seeds	All yellow	6,022 yellow : 2,001 green	3.01 : 1
3. Gray × white seedcoats	All gray	705 gray : 224 white	3.15 : 1
4. Inflated × pinched pods	All inflated	882 inflated : 299 pinched	2.95 : 1
5. Green × yellow pods	All green	428 green : 152 yellow	2.82 : 1
6. Axial × terminal flowers	All axial	651 axial : 207 terminal	3.14 : 1
7. Long × short stems	All long	787 long : 277 short	2.84 : 1

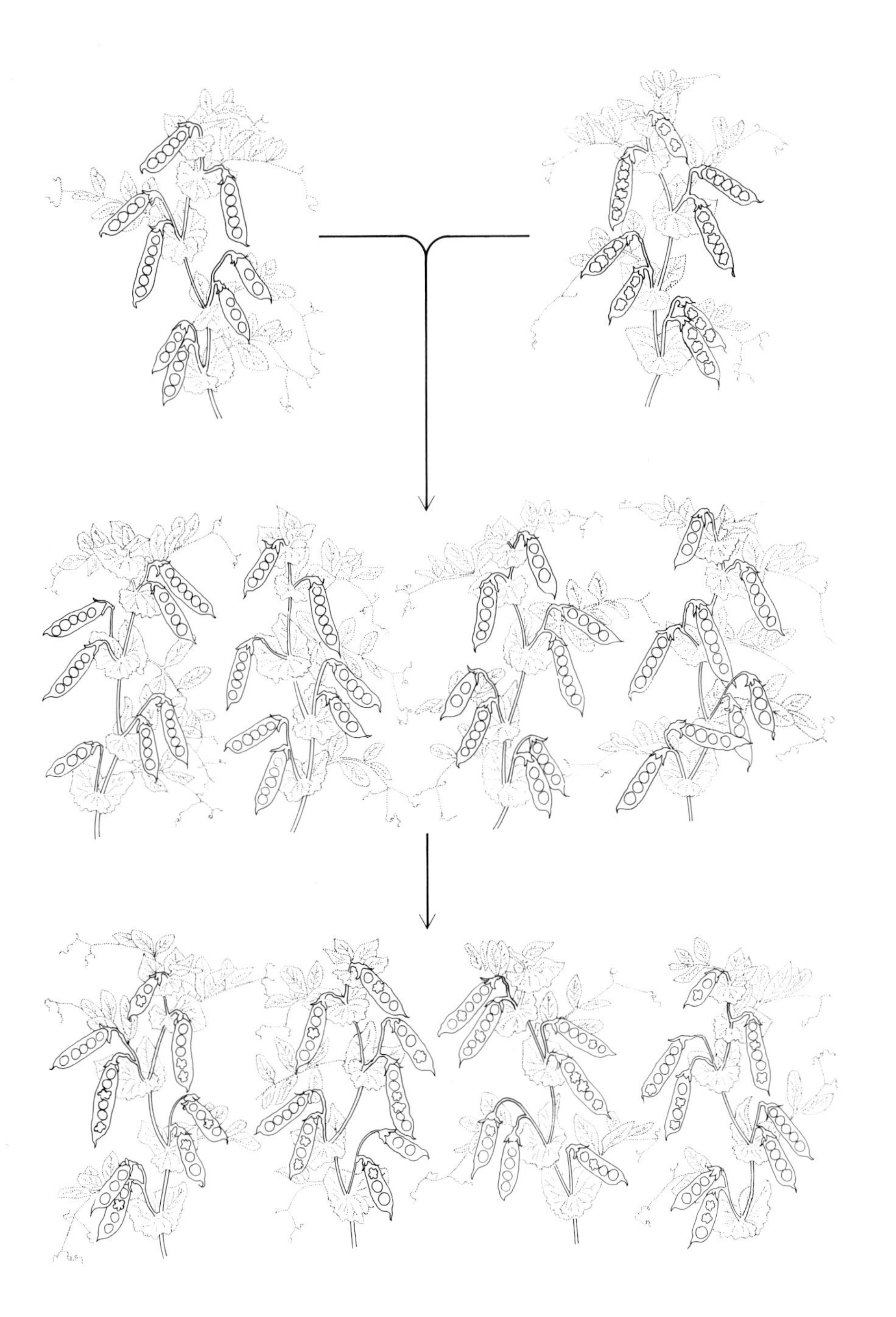

1–3
Mendel found that the wrinkled trait disappeared in the F_1 generation but turned up again in the F_2. (From "The Gene", by Norman H. Horowitz. Copyright © 1956 by Scientific American, Inc. All rights reserved.)

named the trait that disappeared in the F_1, only to reappear in about 25 percent of the F_2, a *recessive trait.*

In order to explain the patterns he observed, Mendel proposed that inherited traits are transmitted from parents to offspring by means of independently inherited "factors" that are now known as *genes.* Furthermore, he found that he could predict the results of his experiments if he assumed that true-breeding lines of plants contributed either a dominant or recessive factor to their offspring in the F_1 generation, and that members of the F_1 were therefore *hybrids.* That is, Mendel postulated that each member of the F_1 contained both dominant and recessive factors. This enterprising monk then invented a shorthand notation by which he could follow his hypothetical dominant and recessive factors through various lineages.

Mendel labeled the factor responsible for the dominant trait (round seeds) *A*, and he designated the factor responsible for the recessive trait (wrinkled seeds) *a*. (Geneticists still use capital letters to represent the genes responsible for dominant traits and lower-case letters to represent those responsible for recessive ones. Which letter is chosen to represent a given pair of alternate traits is arbitrary.) When both parents contribute an *A* to their offspring, the offspring are *AA*, and they produce only round seeds. In *aa* plants, which received an *a* from each parent, only wrinkled seeds are produced. Thus, true-breeding lines of plants are either *AA* (round), or *aa* (wrinkled). What happens if two such lines are crossed, as they were by Mendel to produce the F_1 generation? Clearly, all the offspring receive an *A* from one parent and an *a* from the other, so that all members of the F_1 must be *Aa* with respect to the alternate traits "round" or "wrinkled." What do *Aa* plants look like? Because *A* is dominant to *a*, all individuals in the F_1 will have round seeds. The trait "wrinkled" will seemingly have disappeared from the F_1, just as Mendel observed. But, in fact, the factor responsible for the recessive trait has not disappeared; its effects are simply masked by the presence of the factor *A* and, in later crosses, the effect of *a* can become obvious once again.

Consider what happens when the hybrid plants of the F_1 are allowed to self-fertilize and to produce offspring. All of the parents are *Aa*, and can contribute either *A* or *a* to their offspring, and in fact do so in equal proportions. About half the offspring get *A* and half get *a* from *each* parent. This means that three different kinds of offspring can result: *AA* or *aa* plants if both parents happen to contribute the same factor, and *Aa* plants if each parent happens to contribute a different factor.

An easy way to predict what will happen in a given cross is to construct a table that allows us to keep track of all possible combinations of factors. Across the top of the table are listed the factors that one parent can contribute; those from the other parent are listed down the left-hand side. In our example, both parents are hybrids, so they both can contribute either *A* or *a*, and we represent this as follows:

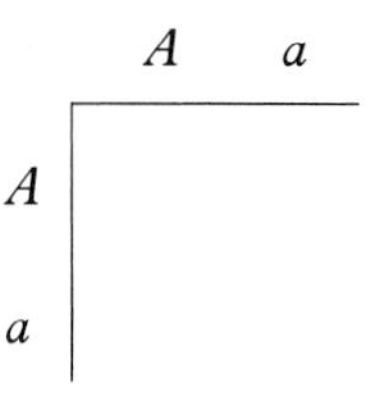

Then by simply drawing in the boxes and combining the factors we generate the following table. (Capital letters are always written first.)

	A	*a*
A	*AA*	*Aa*
a	*Aa*	*aa*

Thus, the combinations we should expect in the offspring are *AA, Aa,* and *aa*. Notice that *Aa* appears in the table twice. This means that about two out of every four offspring will be *Aa*. Or, more generally, about 50 percent of the offspring will be *Aa*. Similarly, about 25 percent of the offspring will be *AA* and the remaining 25 percent, *aa*. Looking at it another way, 75 percent of the offspring are either Aa or AA and therefore have round seeds, and 25 percent are aa and have wrinkled seeds. These ratios are exactly those observed by Mendel in the F_2 generation.

Using the same kind of reasoning Mendel predicted what would happen if he crossed plants differing in *two* alternate characters. He crossed plants that had round yellow peas (both dominant) with plants bearing wrinkled green peas (both recessive) and, as predicted, all members of the F_1 were round and yellow. He then allowed the F_1 hybrids to self-fertilize. If the factors he postulated did indeed exist and behave as independent units, then he expected to find four kinds of peas in the F_2: round yellow, wrinkled yellow, round green, and wrinkled green. Moreover, he predicted he would find them in the ratio 9:3:3:1. He performed the crosses and found the actual ratios to be exactly as he predicted, allowing for small deviations introduced by chance (Figure 1-4).

Mendel had discovered the most fundamental patterns of inheritance, and they have stood the test of time to the present day. We shall soon see how they apply to human lineages. He published his results in 1866, but for the most part his manuscript was ignored. By and large this was because Mendel's "factors" could not be seen; they were rather mysterious units for which no physical basis was known. But all that had changed when Mendel's work was rediscovered in 1900, when it was fully appreciated for the first time. What made the difference was that in the interim biologists had discovered what they presumed to be the physical basis of Mendel's mysterious factors. They had discovered chromosomes.

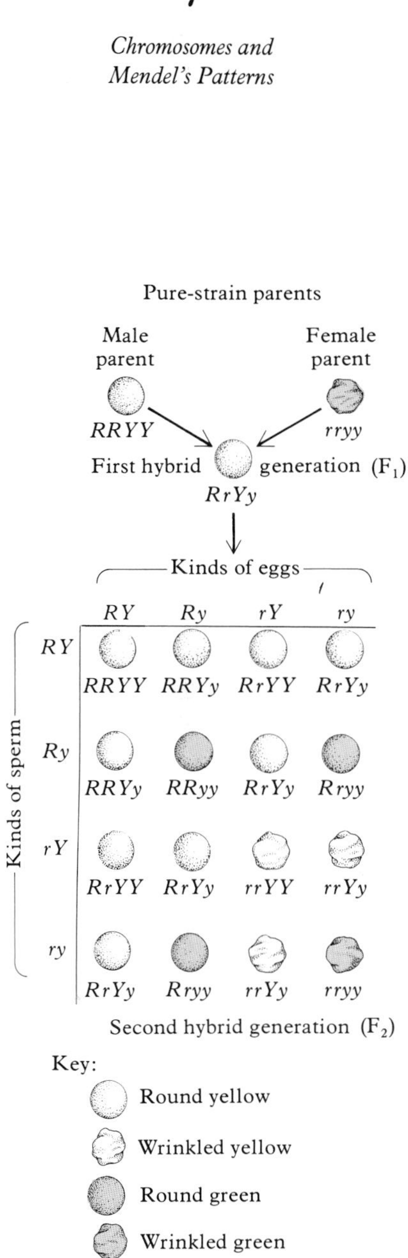

1–4
The results of a cross of pea plants that differ in two *pairs of alternate characteristics (round and yellow versus wrinkled and green).*

Chromosomes and Mendel's Patterns

When Mendel published his results in 1866 it was well known that cells are the basic building blocks of all living things. But at that time these fundamental units were poorly known and largely undescribed, because the manufacturing of microscopes and the preparation of specimens for microscopic study had not yet become highly developed arts. Nonetheless, it was known that most plant and animal cells have a distinct nucleus inside them, and that within the dividing nucleus are rodlike structures called chromosomes. In general, chromosomes are clearly visible only in cells that are in the process of dividing. As we discuss later in this chapter, in resting, nondividing cells, the chromosomes are still inside the nucleus, but they are much thinner and are highly entangled with one another so that individual chromosomes cannot be distinguished. (We should mention here that the relatively simple *prokaryotic* cells of bacteria and blue-green algae lack a distinct nucleus and have a single, unpaired chromosome that is never visible through the light microscope and that is much less complex than the chromosomes of eukaryotic cells that we are discussing here. We will have more to say about the structure of prokaryotic and eukaryotic chromosomes in Chapter 3.)

By the time Mendel's work was rediscovered in 1900 enough was known about the remarkable behavior of chromosomes in dividing cells to suggest that the observed patterns of inheritance could be explained in cellular terms by assuming that Mendel's independent factors were located on chromosomes. The gist of the evidence, as first described about the turn of the century, is this: within the nucleus chromosomes exist in pairs, except in a single, revealing instance. The exception is the sex cells, whose nuclei contain only one member of each pair, or half the number of chromosomes in other body cells. How does this tie in with the inheritance of alternate traits described by Mendel?

Assume that the dominant and recessive forms of a particular factor are located one each on the two chromosomes of a particular pair. Thus, an *Aa* individual has an *A* on one chromosome of a given pair and an *a* on the other. When the individual produces sex cells, pairs of chromosomes separate so that each egg or sperm contains either an *A* chromosome or an *a* chromosome. Then, when self-fertilization occurs, pairs of chromosomes are reunited once again, and the resulting offspring are either *AA, Aa,* or *aa,* depending on which factors happened to be located on the particular chromosome pairs that were reunited.

You will recall that Mendel followed the patterns of inheritance of seven pairs of alternate traits of the common garden pea. It turns out that pea plants have seven pairs of chromosomes, which correlates with Mendel's observation that all seven of the traits he studied behaved independently. That is, Mendel's factors showed *independent assortment* because virtually all of the traits he studied are determined by factors located on different pairs of chromosomes. (Mendel was thus not only careful, but lucky, too.)

It is now known that the number of chromosome pairs normally present in the nucleus can vary widely from species to species. (Remember that only one member of each pair is present in an animal's sex cells.) It is also known that each chromosome pair usually carries factors responsible for many different traits.

How does all of this relate to people? Our discussion of Mendel's work provides a background for discussing human chromosomes and for relating them to the patterns of inheritance shown by some alternate traits in human families. But before we go any further we should first introduce some terms that describe the genetic make-up of an individual, human or otherwise. Familiarize yourself with these words now, for they will be used repeatedly in the discussion that follows.

Some Definitions

Mendel's inherited factors, the units of heredity, are now called *genes.* We will discuss the biochemical basis of genes in following chapters. The two (or more) forms of genes responsible for alternate traits (*A* and *a* in our example) are called *alleles.*

Individuals in whom the two alleles of a given pair are the same (*AA* or *aa*) are called *homozygotes,* whereas *heterozygotes* are individuals in whom the two alleles of a given pair are different (*Aa*).

Recall that you cannot distinguish *Aa* heterozygotes from *AA* homozygotes merely by looking at them. (Both have round seeds.) But the two can be told apart if they are crossed with known heterozygotes (*Aa*). Thus, if wrinkled seeds (*aa*) turn up in the offspring, we can conclude that both parents must have been *Aa.* If only round seeds are produced, then the offspring are either *Aa* or *AA,* and the parent crossed with the known heterozygote must have been *AA.* To distinguish individuals that look alike but nonetheless have different genetic constitutions geneticists use the terms *phenotype* and *genotype.* Phenotype is a description of what an individual looks like, and genotype describes the individual's genetic constitution. In our example, individuals of phenotype "round" may be either of two genotypes, *Aa* and *AA.*

With these terms in mind, let us discuss the chromosomes of the human species and relate them to some fairly obvious patterns of inheritance in human families.

Human Chromosomes

Although the existence of chromosomes has been known for over a hundred years, the exact number of chromosomes that characterizes the human species was not discovered until 1956. This seems extraordinary, but in fact, the chromosomes of most mammals are not only numerous but difficult to prepare for detailed microscopic study; only relatively recently have satisfactory and consistent methods for visualizing them been developed. Human cells contain a total of 46 chromosomes in 23 pairs. That is, body cells, or *somatic cells,* contain 46 chromosomes in their nuclei, and sex cells contain only a single member of each pair, or a total of 23.

Human chromosomes can be made visible for detailed study in the following way. First, a few cubic centimeters of blood are collected in a syringe and then inoculated into a culture dish containing chemicals that stimulate some white blood cells in the sample to divide, thereby making their chromosomes visible. When the white cells have started to divide, other chemicals are added to make the chromosomes swell and to stop the process of cell division at a stage when the chromosomes are most easily distinguished from one another. The cells are then broken open and examined under the microscope. The result, as shown in Figure 1-5, for a blood sample obtained from a normal human male,

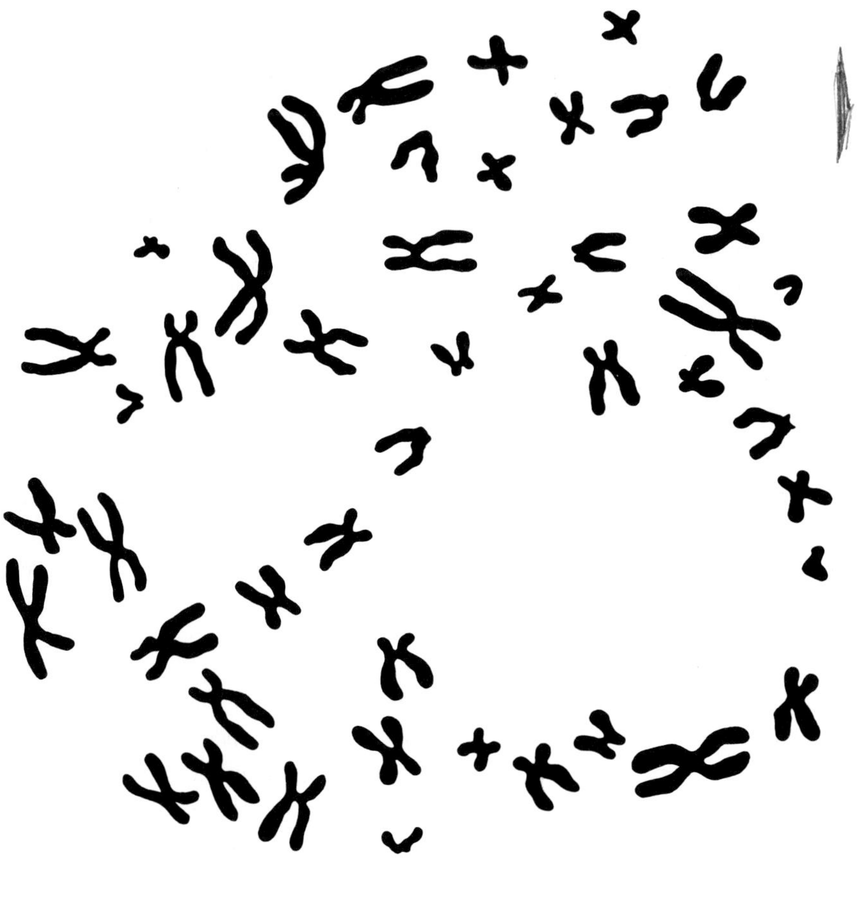

1–5
A chaotic array of chromosomes of a normal human male.

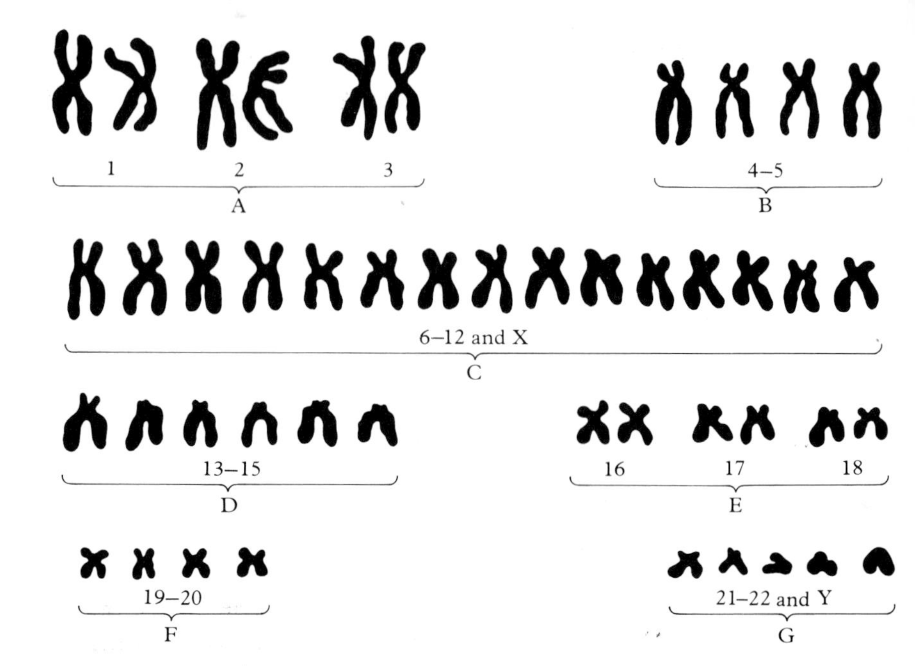

1-6
The karyotype of a normal human male. The chromosomes have been photographed, cut out, and arranged in groups according to size and shape.

is a distinct but chaotic array of human chromosomes. These are then photographed and individual chromosomes are cut out of the photo like paper dolls and lined up with one another in matching pairs to form a *karyotype,* as shown in Figure 1-6. (More recently, computers have been used for the matching process.)

The 23 pairs of human chromosomes are each placed in one of seven groups designated by the letters A through G (Figure 1-6). Of the 23 pairs, 22 are for all practical purposes perfectly matched in both sexes and are called *autosomes.* The remaining pair are called *sex chromosomes,* and though the members of this pair are apparently identical in women, they are not identical in men. With regard to genotype, women are said to be of sex chromosome constitution *XX,* whereas men are said to be *XY.* We will discuss sex chromosomes further in the following chapter. For now, we turn our attention to the inheritance of some human traits determined by genes located on autosomes. But first, it is helpful to elaborate a little on exactly how the terms dominant and recessive apply to human characteristics.

Most of what is known of human genetics has been learned by studying various kinds of diseases or abnormalities that obviously run in families. In the final analysis, all gene-dependent differences among human beings are differences between physiological processes that occur inside cells. Sometimes the way in which genetic differences between cells can result in phenotypic differences between individuals is obvious. For example, albinos are lightly pigmented because their cells are unable to properly synthesize the dark-colored pigment melanin. But oftentimes it is not at all obvious how phenotypic abnormalities relate to gene-dependent abnormalities in cellular physiology and in biochemistry. For example, it is not obvious how the gene for six-fingered dwarfism (see the following discussion) brings about its effect. Moreover, the exact biochemical or physiological defect that is associated with a genetically determined abnormality is known for only about one trait in five.

Many heritable human disorders, most of them individually rare, are the result of a single abnormal allele for which an affected person may be either homozygous or heterozygous. These rare disorders therefore have simple Mendelian patterns of inheritance, as we are about to discuss. An abnormal allele is dominant or recessive depending on whether its effects are evident in a single dose (in heterozygotes) or whether the allele must be present in a double dose (in homozygotes) to produce its effects.

On the other hand, many common disorders, such as high blood pressure and some relatively frequent congenital malformations such as cleft lip and palate, do not have simple Mendelian patterns of inheritance. This is because these abnormalities, like many others, result from the interaction of many genes and many nongenetic environmental factors. As we will discuss in Chapter 5, this is also true for many "normal" human characteristics such as height and intelligence. In the discussion of the patterns of inheritance shown by autosomal dominant and recessive traits that follows, it should be borne in mind that the existence of rare abnormal alleles in affected persons implies the presence of normal alleles in normal people. The study of genetic abnormalities can thus help us to get an idea of how extensive the normal human genetic program really is.

Autosomal Dominant Inheritance

About 450 human traits are known to have their genetic basis in dominant genes located on autosomes. As you know, Mendel discovered that dominant traits are manifested both by heterozygotes (*Aa*) and by homozygotes (*AA*). But almost all human beings who manifest documented autosomal dominant traits turn out to be heterozygotes. This is because dominant genes for the most part produce undesirable effects. That is, persons manifesting dominant traits are usually at some kind of disadvantage compared to their normal peers. Apparently, homozygotes for dominant traits are at such a disadvantage that most of them do not survive life before birth and die as embryos. Besides, autosomal dominant traits are rare to begin with, so it is unlikely that two affected persons would come together to produce homozygous offspring. Thus, we can usually assume that persons manifesting autosomal dominant traits are heterozygous.

A good example of autosomal dominant inheritance is provided by the rather benign trait known as "wooly hair," whose distribution has been well documented in Norwegian families. Affected persons have hair that is tightly kinked and very brittle, so that it breaks off before growing very long (Figure 1-7). As usual, people manifesting this dominant trait are heterozygous, and

1–7
A Norwegian family, some of whose members have woolly hair. (From Mohr, Journal of Heredity, 23, *1932.)*

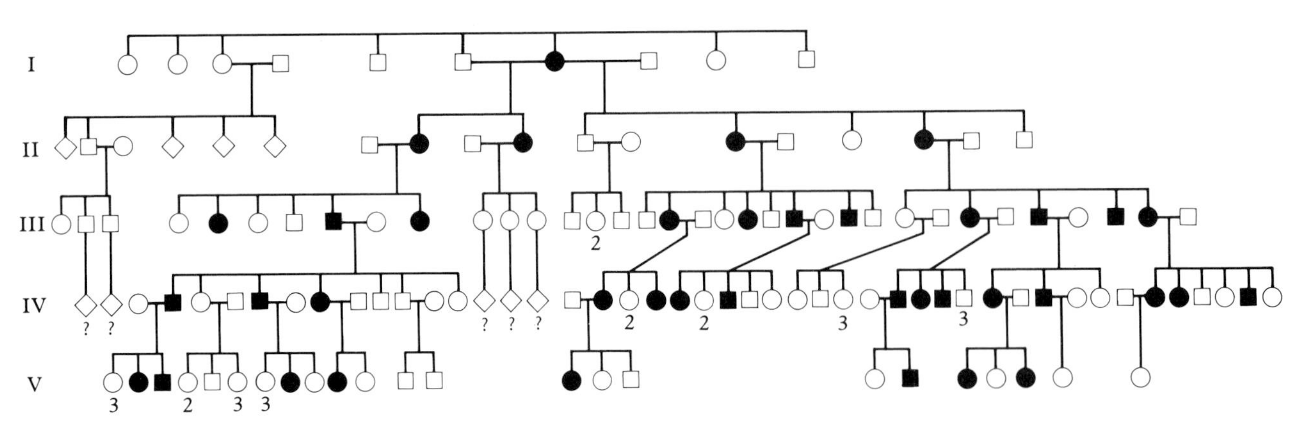

1–8
A pedigree showing the transmission of woolly hair through five generations. If more than one offspring are represented by a symbol, the number represented is given below the symbol. (After Mohr, Journal of Heredity, 23, *1932.)*

their genotype can be symbolized *Ww* (*W* for wooly). If an affected person (*Ww*) and an unaffected, "normal" person (*ww*) produce offspring, we would expect about half of them to have normal hair and half to have wooly hair. Moreover, if the gene that determines the trait is located on an autosome, then the affected offspring should include roughly equal numbers of males and females, and the trait should be transmitted from either parent to both sons and daughters. That this is true is shown in Figure 1-8, which is a human pedigree outlining the transmission of wooly hair through several generations. The symbols used in outlining pedigrees are these: women are represented by circles and men by squares. Matings occur between individuals directly connected by horizontal lines, and their offspring are indicated at the end of short vertical lines. Affected individuals are indicated by blacking in the symbols. For example, suppose a normal man marries a wooly-haired woman and that they produce four offspring, two boys and two girls, one affected and one normal child of each sex. This is symbolized as follows:

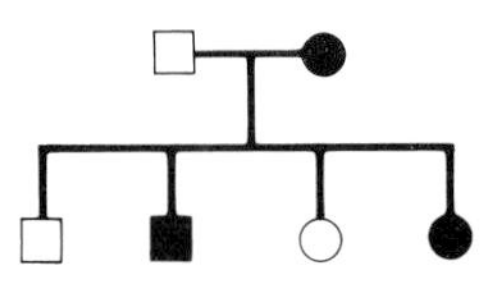

If the affected daughter then marries a normal man and produces five offspring, including three normal daughters and two affected sons, the diagram is extended to:

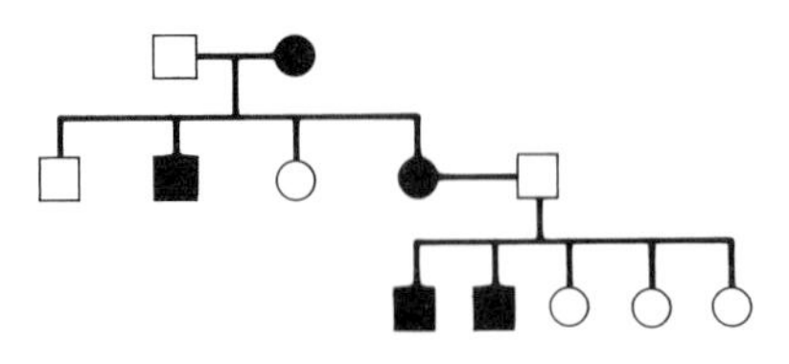

Pedigrees of families including wooly-haired individuals have also been recorded in Holland and the United States. If we pool all the data concerning the offspring from marriages between one affected and one normal person, we find, as we would expect, that within the limits of chance half of the sons and half of the daughters inherit the trait.

A fairly constant feature of most autosomal dominant traits in human families is that they can vary widely in severity from one individual to another.

The degree of severity is referred to as the *expressivity* of a particular trait. For example, the condition known as Marfan's syndrome is an autosomal dominant disorder of connective tissue that is manifested by abnormalities of the position of the lens of the eye, by excessively long bones in the hands, feet, and extremities, and by defects in the wall of the aorta as it comes off the heart. People who have Marfan's syndrome can be anywhere from severely affected with all of these abnormalities to slightly affected with only abnormally long fingers to reveal the presence of the disorder. (In fact, because of his distinctive appearance and family background, it has been suggested that Abraham Lincoln may have been mildly afflicted with Marfan's syndrome. See Figure 1-9.)

By and large, whether or not a particular autosomal dominant gene is fully expressed depends on the rest of the person's genes. In other words, the presence of certain other genes or combinations of genes can markedly influence the expressivity of a genetically determined trait, particularly of autosomal dominant ones.

Another feature of autosomal dominant traits is that they sometimes appear unexpectedly among the offspring of unaffected parents. In Marfan's syndrome, this occurs about 15 percent of the time. Those who are affected thereafter pass the trait on to about half of their offspring, as expected. In general, the sudden, unexpected appearance of an autosomal dominant trait in a lineage from which it was not previously known is the result of a *mutation.* In our example, a heritable change occurs spontaneously within the genetic material and thereby transforms a normal gene into the one responsible for Marfan's syndrome. (We will discuss mutations further in following chapters.)

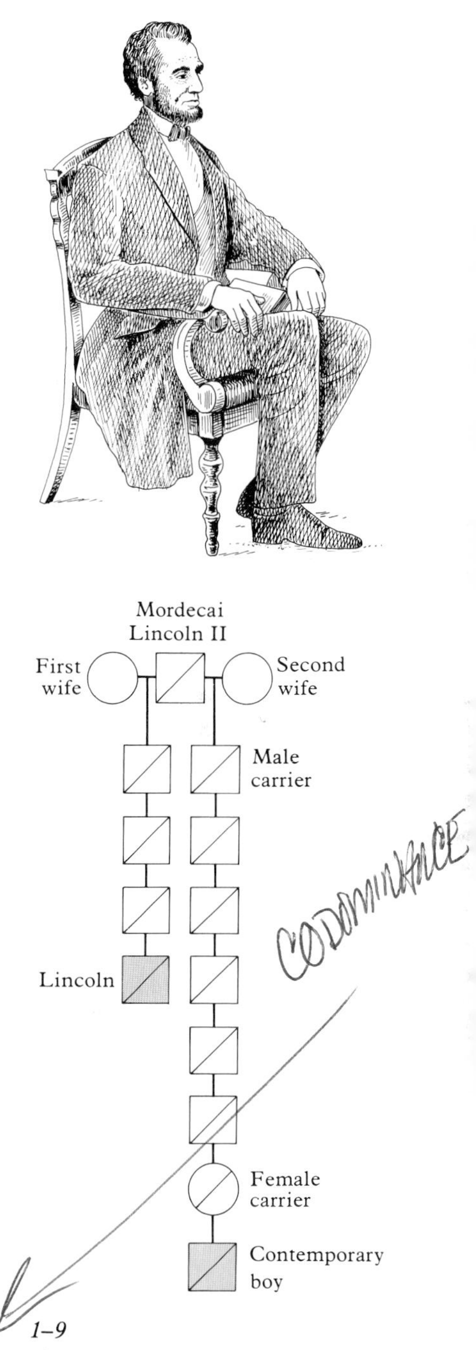

1–9
Abraham Lincoln's extremely long legs and the very unequal lengths of his thumbs may indicate that he was mildly afflicted by Marfan's syndrome. As shown in the pedigree, this disease was recently diagnosed in a boy who is a distant relative of Lincoln's. The boy and the famous president are both descendants of Mordecai Lincoln, II, but they had different mothers. Mordecai almost certainly had the gene for Marfan's syndrome, but he showed no obvious symptoms because in him the gene had a low expressivity.

Codominance, The ABO Blood Group

So far we have been discussing autosomal dominant traits that depend on a single pair of alleles, which is to say that the genes determining such traits come in only two forms, *A* and *a.* As we have seen, most people who manifest autosomal dominant traits are heterozygous for the allele in question. However, *multiple alleles* are also known to influence the inheritance of some well-known and important human characteristics, including the determination of a person's *blood group.* (There are many blood groups, most of which depend on different sets of alleles.) Perhaps the best known set of multiple alleles is that determining the ABO blood group, which is important enough to be worth discussing further.

Three alleles determine to which ABO blood group a person belongs. These can be symbolized: I^A, I^B, and I^O. These alleles are always found at a particular place, or *locus,* on a particular pair of autosomes, and any one person has two out of the three alleles. I^AI^A individuals are of blood group A, as are I^AI^O individuals. Similarly, persons who are I^BI^B or I^BI^O are group B. To complete the possibilities, I^AI^B individuals are of blood group AB, and I^OI^Os are of group O.

Both of the alleles I^A and I^B are dominant to I^O. Moreover, when I^A and I^B occur together, both have an effect on the surface of red blood cells (and both therefore stimulate the production of antibodies). Thus I^A and I^B are said to show *codominance.*

ABO blood grouping has been carried out nearly world wide for several reasons. First, ABO compatibility is essential in performing blood transfusions. And second, ABO groups vary enough from one group of people to another that their patterns of variation have been useful to physical anthropologists in the study of human races (see Chapter 4).

An unintended but useful side effect of widespread ABO typing has been its application to cases of disputed paternity. Consider the following example. During the course of divorce proceedings a woman of Type A (genotype I^AI^A or I^AI^O) seeks child support from a man of Type O (genotype I^OI^O), claiming that he is the father of her recently born child. If it should turn out that the child is of Type AB (genotype I^AI^B), the case would be thrown out of court, because the supposed father is capable of contributing only the allele I^O to his offspring. The mother can contribute only I^A or I^O, so the allele I^B must have been contributed by someone else. (The likelihood of a mutation having occurred in one of the parents' sex cells is negligible.) However, if the presumed father were of Type B (genotype I^BI^B or I^BI^O) then overall there would be a 50 percent chance that he could in fact be the father and the trial would continue.

We now turn our attention to the inheritance of some autosomal traits whose expression implies that an affected person, unlike most people who are afflicted by a dominant trait, is a homozygote.

1–10
Two albino parents and their albino daughter. (After Davenport, Journal of Heredity.*)*

Autosomal Recessive Inheritance

At least 500 human traits are definitely known to show an autosomal recessive pattern of inheritance. As first discovered by Mendel, heterozygous individuals (*Aa*) do not manifest autosomal recessive traits because of the masking effect of the dominant allele. Thus, people who manifest autosomal recessive traits are generally *homozygous recessive;* that is, their genotype is *aa.*

As compared with their dominant counterparts, autosomal recessive traits are likewise usually associated with diseases or abnormalities, and they are not quite so rare. This is because most people who are heterozygous for an autosomal recessive trait (of genotype *Aa,* also called carriers of the trait) are not at much of a disadvantage compared to normal (*AA*) individuals, so the trait may become widely disseminated, even if affected homozygous people choose not to reproduce. Also, those affected by autosomal recessive traits tend to be less variable than are those affected by autosomal dominant ones. That is, the expressivity of autosomal recessive traits is about the same for all those who are affected.

A good example of autosomal recessive inheritance is the condition known as *albinism.* Albinism is one of the most common and widespread of genetic disorders. Affected individuals include not only human beings of all races, but also other mammals, insects, fish, reptiles, amphibians, and birds. Albinism results from the body's inability to properly synthesize the dark-colored pigment *melanin.* Melanin is the principal pigment that imparts color to human skin, hair, and eyes, so human albinos generally have white hair and pink or only lightly colored irises (Figure 1-10). Because of their lack of pigment, the skins and eyes of albinos are abnormally sensitive to the effects of sunlight, and because of their unusual appearance human albinos may receive special treatment from other members of their species. For example, the Aztec emperor Montezuma is said to have included many albinos among the members of his "museum" of living human "curiosities," and albinos among the present-day San Blaz Indians of Panama (known as "moon children" because they avoid bright sunlight) are not permitted to marry.

Until quite recently it was thought that albinism was the result of a single, specific defect in the synthesis of melanin from the amino acid tyrosine. (In later chapters we will discuss the biochemistry of melanin synthesis as it relates to inborn errors of metabolism.) But then a well-documented pedigree of albinism in England showed that two albino parents produced four children, none of whom were albinos. As shown in Figure 1-11, the mating of parents manifesting the same autosomal recessive trait can produce only affected offspring, because each of the parents contributes an *a*. How then can the pattern observed in this unusual English family be explained?

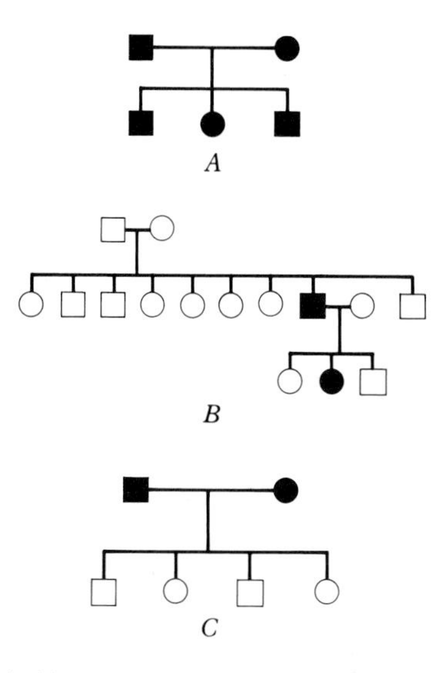

1–11
A, *The mating of albino parents almost always results in all albino offspring.* B, *A pedigree showing albinism among the offspring of unaffected carrier parents and among the offspring of an albino father and an unaffected carrier mother.* C, *A pedigree of albinism in which albino parents produced four nonalbino offspring.*

It turns out that there are at least two, and perhaps as many as six, genes that result in albinism, depending on where the biochemical block in the synthesis of melanin occurs. And all of these defects in melanin synthesis are inherited as autosomal recessive traits. Therefore, it is possible for two albinos who are homozygous recessive for different defects in melanin synthesis (and who are therefore albinos for different reasons) to produce normal offspring, as shown in Figure 1-11.

In the United States, about one white person in 38,000 and one black person in 22,000 are albinos. But circumstances are known under which the percentage of albino offspring produced is higher than in the population at large. The best known example is that of marriages between relatives, or *consanguineous* marriages. Although only about 0.1 percent of marriages in the United States are between first cousins, about 8 percent of albino children result from first-cousin marriages. (First cousins are the offspring of brothers and sisters who married unrelated spouses. See Figure 1-12.) How does the incidence of albinism, or any other autosomal recessive trait, relate to the degree to which an affected person's parents are related?

Most autosomal recessive traits are rare. Nonetheless, the brothers and sisters of someone who carries an allele for a rare autosomal recessive trait are very likely to be carriers too. (If a person carries an allele for albinism, the person's normal brothers and sisters have a 50 percent chance of also carrying the allele, because at least one of their parents must be a heterozygote. Work this out for yourself.) Similarly, people descended from a common ancestor known to have manifested or carried a particular trait are also more likely to be carriers. Thus, people manifesting rare autosomal recessive alleles tend to cluster in certain family lines because the mating of rather closely related individuals is likely to bring two rare autosomal recessive alleles together to produce an affected individual.

One example is the autosomal recessive condition known as six-fingered dwarfism, which is unusually frequent among the Old Order Amish of Lancaster County, Pennsylvania (Figure 1-13). Amish people usually choose to marry other Amish people, and because their numbers are relatively small to begin with, this means that marriages between individuals who have common ancestors occur frequently. Accordingly, whereas six-fingered dwarfs are very rare elsewhere, they exist in at least 33 Amish families. Apparently, one of the original founders of the sect in eastern Pennsylvania was a heterozygous carrier of the gene for six-fingered dwarfism: the gene has since become widespread in the population and is frequently found in the homozygous state because of consanguineous marriages.

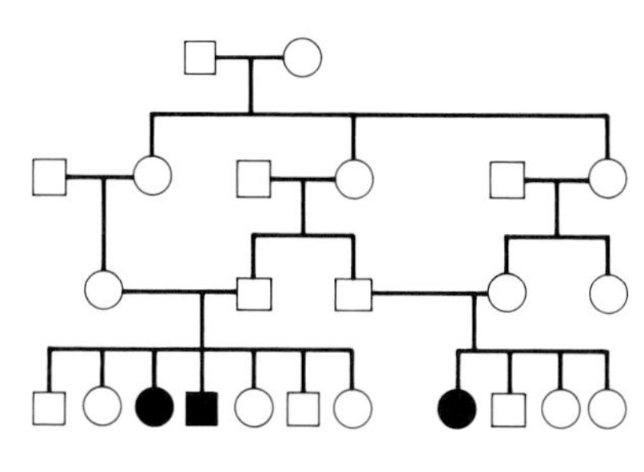

1–12
A pedigree showing cousin marriages that resulted in albino offspring. The consanguineous matings occurred in the third generation.

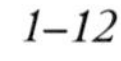

In general, the more closely related two people are, the greater the chance that their offspring will manifest some (usually detrimental) autosomal recessive trait. This is best illustrated by data about the offspring of incestuous unions of fathers and daughters and brothers and sisters. Information on the offspring of

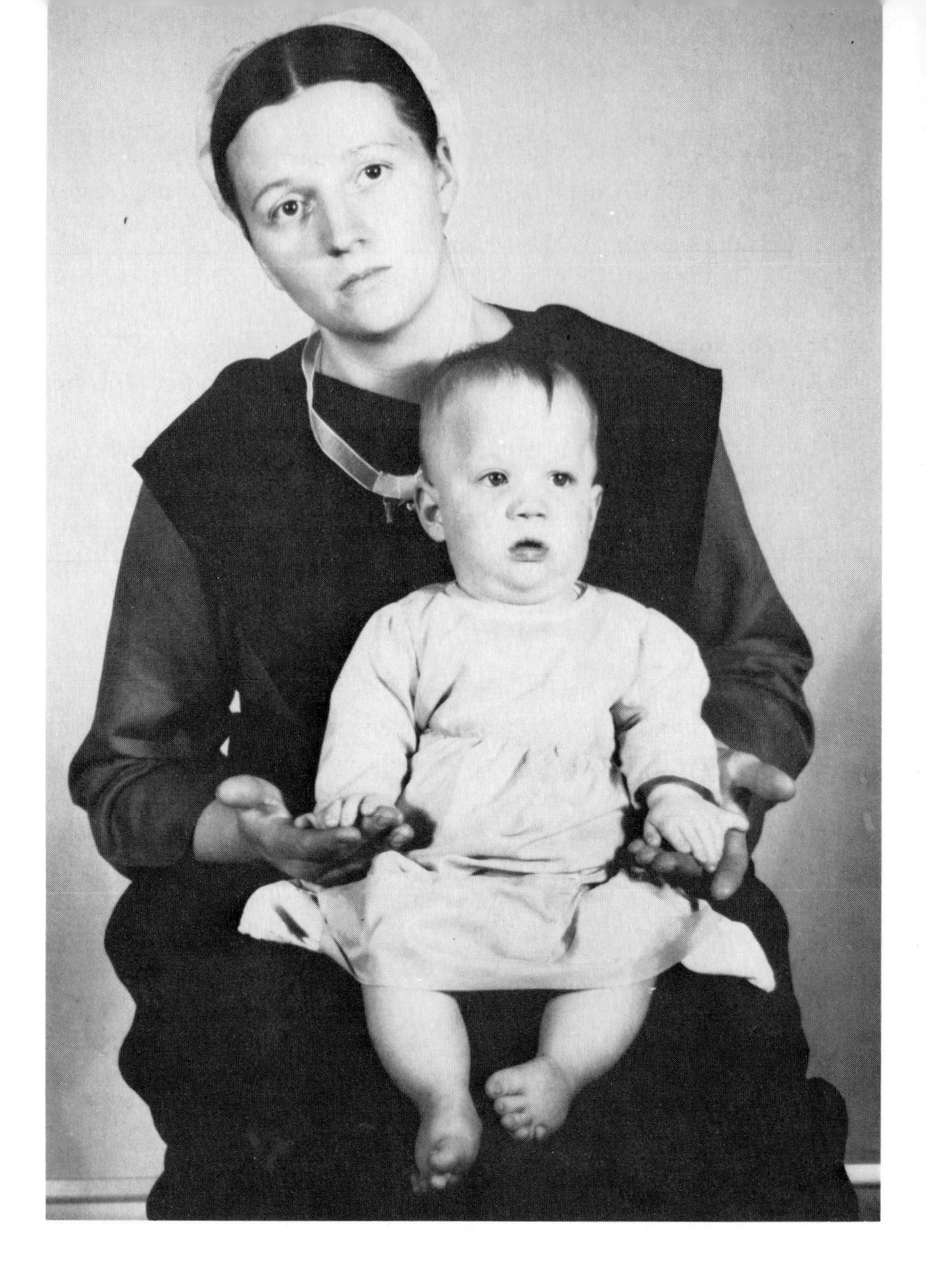

1–13
An Amish mother and her child, who has six-fingered dwarfism. (Photo courtesy of Dr. Victor A. McKusick.)

31 such unions is available from England and the United States. Six of the offspring died early in life and twelve were severely affected physically or severely retarded mentally. Only 42 percent of the children were apparently normal. (Almost all human societies have strict cultural prohibitions against incestuous unions. Is this because the deleterious genetic consequences of such unions are so obvious? Almost certainly not. Rather, it appears that incest taboos are outgrowths of social and cultural factors, and not the result of a primitive form of applied genetics.)

Autosomal recessive traits are of particular interest in genetic counseling. Most often the relatives of an affected person seek advice about whether they are carriers, or about the chances of their having an affected child. It is sometimes possible to infer a person's genotype directly from pedigree studies, and the person can thereby be told his or her status as a carrier. But even when a person's genotype cannot be deduced by pedigree analysis, it is often possible to determine it by performing various physiologic and biochemical tests. For some heterozygotes for autosomal recessive traits do indeed manifest the fact that they harbor both dominant and recessive alleles.

We have already discussed the fact that homozygous individuals who manifest autosomal recessive traits usually do so about equally. The effects of a dominant allele generally mask the effects of a recessive one in heterozygotes, but in some instances it is possible to infer the presence of the recessive allele by subjecting the suspected carrier to some kind of stress. For example, heterozygous carriers for sickle-cell anemia (whose biochemistry and genetics we will discuss further in Chapter 3) can be detected if their blood is subjected to lower-than-normal concentrations of oxygen. Under this stress, some of the carrier's red blood cells take on the characteristic sickle shape, and thus reveal heterozygosity.

In general, it is possible to devise some sort of measurement that will reveal heterozygous carriers of autosomal recessive traits, as long as the physiological or biochemical defect in question is known. But the exact defect is known for only about one genetically determined abnormality in five. Besides, the inheritance of many traits of normal human beings does not fit into any of Mendel's patterns, in spite of the fact that the traits obviously have some sort of genetic basis. This is especially true of traits that are not clearly alternate, but rather are continuously distributed throughout the population. Such traits do not manifest themselves as sharply defined pairs of phenotypes like "round" or "wrinkled," but rather as continuous gradations. For example, normal body height varies widely but in general is continuously distributed in a given population (Figure 1-14). Even though there is a definite tendency toward tallness or shortness in families, there may be widespread differences in the heights of parents and offspring. Also, as with most continuously varying traits, height is known to be affected, not only by a person's genes, but by the environmental conditions in which the person grows up and lives.

It was Mendel himself who first proposed an explanation for the inheritance of continuously graded traits. Based on some experiments he performed on bean plants with colored and white flowers, he suggested that *more than one pair of genes* transmitted such traits. For the most part, Mendel's explanation has stood the test of time, though we now are much more aware of the effects of the environment on the expression of continuously varying traits than Mendel was. We will return to the inheritance of continuously varying traits and to the effects of the environment on the expression of a person's genotype in later chapters. But for now, let us return to our discussion of autosomal dominant and recessive traits and ask a deceptively simple question. Which genes are located on which chromosomes?

1–14
A company of 175 soldiers arranged in groups according to height. The lower row of numbers indicates height in feet and inches, the upper row the number of men in each group. (From Blakeslee Journal of Heredity, 5, *1914.)*

Mapping Human Chromosomes

About 1200 human genes are definitely known to exist, and new ones responsible for normal or abnormal traits are discovered each year. It has been estimated that the total number of genes per average human being will probably eventually be measured in thousands or perhaps tens or even hundreds of thousands, but nobody knows for sure. At any rate, there are clearly thousands more genes than chromosomes, which means that in general each chromosome contains a large number of genes. Parents contribute entire chromosomes to their offspring, so we would expect all of the genes on a given chromosome to be inherited *en masse,* and such collections of genes make up a chromosomal *linkage group.* Thus, for human beings we would expect 23 groups of linked genes corresponding to the 23 pairs of chromosomes. This will probably turn out to be true, but our present knowledge of human linkage groups is meager. Only one human chromosomal linkage group, that associated with sex chromosomes, has been described in any detail, as will be discussed in the following chapter. One of the major reasons linkage groups are so difficult to figure out is that most of the time the linkage of genes on any chromosome is not complete. What this means is well illustrated by the test cross that first demonstrated the existence of linkage in the early 1900s.

You will recall that the seven traits of the common garden pea studied by Mendel showed independent assortment because the genes responsible for the traits are located on different pairs of chromosomes. But some experiments on sweet peas by later investigators suggested that not all traits are inherited independently. True-breeding sweet peas that had red flowers and spherical pollen grains (both dominant) were crossed with plants that bore purple flowers and had cylindrical pollen grains (both recessive). If these traits were determined by factors that showed independent assortment, the offspring of crosses of members of the F_1 generation with individuals who were homozygous recessive for both traits were expected to be of four different phenotypes in the ratio 1:1:1:1. But to the investigators' surprise the actual ratios turned out to be 7:7:1:1. Furthermore, the two smaller categories of offspring manifested combinations of traits that were not found in the parents. The explanation offered was that combinations of traits that occurred together almost all of the time did so because the factors responsible for them were located on the same chromosome. But what about the unexpected *recombinations* of traits observed in the offspring? Further experiments revealed a remarkable fact: the members of each pair of chromosomes sometimes physically exchange sections with one another, thus unlinking traits that are found together on the same chromosome and allowing for the generation of new combinations. This exchange occurs when sex cells are produced during meiosis, and the exchange of sections between chromosomes of a given pair is known as *crossing over.*

Meiosis is a process in which cells divide in such a way as to reduce the number of chromosomes in the nucleus by half. Meiosis is thus often referred to as *reduction division,* and in general it applies only to the production of sex cells. On the other hand, when a somatic cell divides, the two cells that are produced have the same number of chromosomes as the original cell before it divided. This is because somatic cells duplicate their entire set of chromosomes before they divide and then distribute a complete set to each of the two cells produced by division. (This type of cell division is called *mitosis.*) But in meiosis, one cell divides twice and produces four cells, each of which has only half of a com-

plete set of chromosomes, one member of each chromosome pair. How this occurs is shown in Figure 1-15.

The most important features of meiosis as it relates to crossing over and linkage can be summed up in the following way. Like a somatic cell, a cell undergoing meiosis duplicates its entire set of chromosomes before it divides the first time. But before the duplicated sets of chromosomes separate in the first division of meiosis they do something generally unheard of among the chromosomes of somatic cells. Before the first division most chromosomes

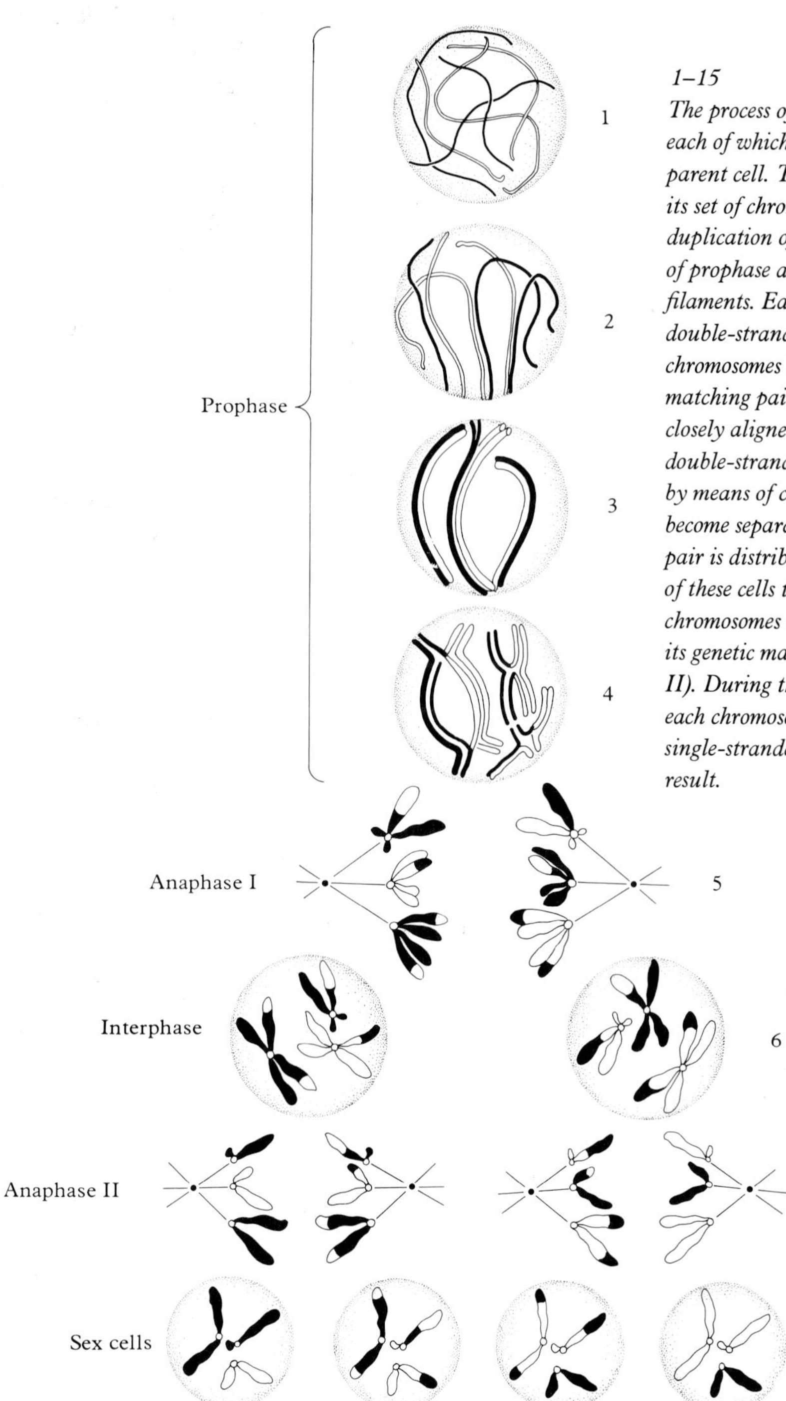

1–15
The process of meiosis results in the formation of four cells, each of which has half the number of chromosomes as the parent cell. The parent cell accomplishes this by duplicating its set of chromosomes once and then dividing twice. The duplication of the chromosomes occurs just before the onset of prophase and they first become visible as thread-like filaments. Each duplicated chromosome is actually double-stranded, and this becomes more apparent as the chromosomes become shorter and thicker. During prophase matching pairs of double-stranded chromosomes become closely aligned with one another and segments of the double-stranded matching pairs may be physically exchanged by means of crossing over. During anaphase I the pairs become separated. One double-stranded member of each pair is distributed to each of the two cells that result. Each of these cells thus contains half as many double-stranded chromosomes as the parent cell. Then, without duplicating its genetic material, each of these cells divides again (anaphase II). During the second division the two strands that make up each chromosome separate and each strand itself becomes a single-stranded chromosome in one of the four cells that result.

undergoing meiosis become closely aligned with one another in matching pairs along their entire lengths, as shown in Figure 1-15. And it is during this time of close alignment that the physical exchange of sections of matching pairs of chromosomes becomes apparent. As they separate during the first division, paired meiotic chromosomes are connected to one another at X-shaped areas called *chiasmata* (singular, *chiasm*), which presumably are points at which crossing over has occurred. Then, after the duplicate pairs have fully separated and formed two cells, each newly formed cell divides again to produce four cells, each of which has half of a complete set of chromosomes.

By about 1910 or so it was known that genes are strung out on a chromosome like beads on a string. But the exact nature of the relationship between genes and chromosomes and of the physical and biochemical events of crossing over between duplicated sets of chromosomes remain largely unknown. Whatever its physical basis, crossing over is of enormous importance in helping to produce the genetic variability that is the raw material of evolution. This is because crossing over generates an enormous number of combinations of genes both in eggs and in sperm cells, and thus helps to provide the raw material on which natural selection can operate.

In general, how much crossing over occurs between genes on the same pair of chromosomes depends on the location of the genes with respect to one another. Genes farther away from one another cross over more often than those closer together. This observation provides a basis for constructing *chromosome maps.* Thus, if crossing over between two genes occurs about 1 percent of the time, then the two are separated by "1 map unit" of distance. The further away the two genes are, the greater the number of crossovers. In fact, if two genes are separated from one another by more than 50 map units, crossovers may be so frequent as to suggest that the genes are not linked, but located on different chromosomes (Figure 1-16).

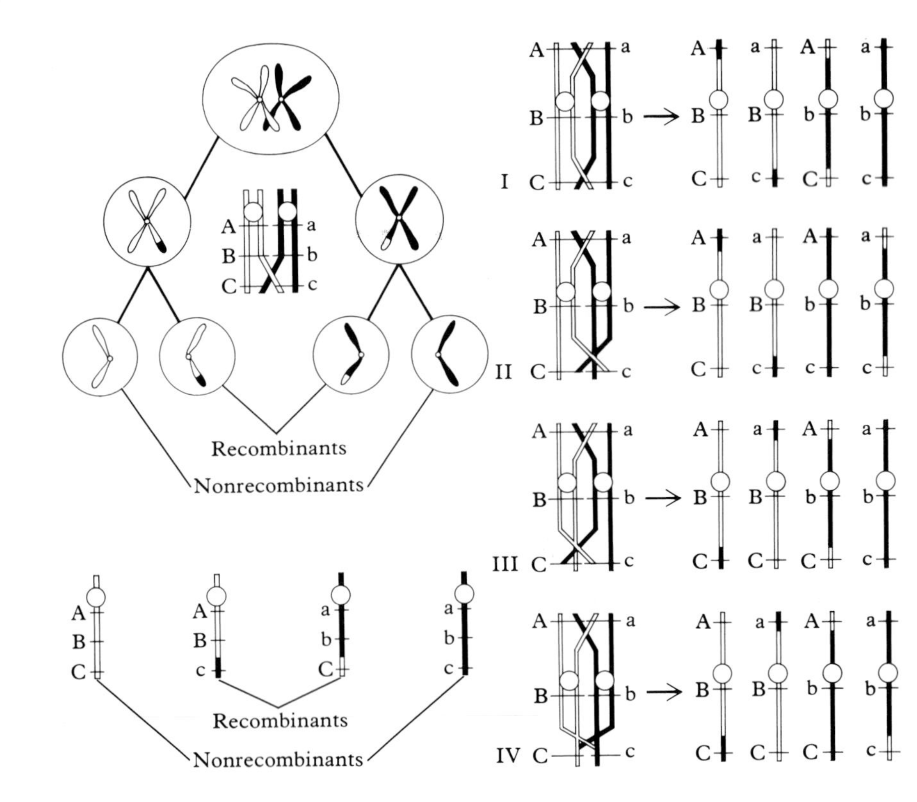

1–16
Segments of matching pairs of chromosomes are physically exchanged by crossing over. Each member of the pair of chromosomes shown here carries three alleles. In crossing over, corresponding segments of the arms of the chromosomes can be physically exchanged (left.). This results in the four chromosomes shown at the bottom left, two of which have a different combination of alleles than was present in either of the original chromosomes. Right, the four kinds of recombinations for three dominant and three recessive alleles.

Extensive chromosome maps have been worked out for a few experimental animals, including mice and fruit flies. (Nearly complete maps are also known of the single, circular chromosome of some prokaryotes, particularly bacteria. See Chapter 3.) But our ignorance of the human map is truly profound. It is estimated that less than 5 percent of the human chromosome map is known. This amounts to a total of some 200 or so human genes whose chromosomal locations have been assigned. And of the genes mapped so far more than half are located on a single chromosome, the sex chromosome known as the X-chromosome. The entire X-chromosome probably has about as many genes as an autosome of similar length. But as we shall discuss in the following chapter, it is relatively easy to tell that a trait is on the X-chromosome because of the characteristic pattern of inheritance of X-linked traits. This is clearly not true for autosomes. Although it is sometimes easy to determine whether a given trait is inherited as an autosomal dominant or recessive, it is usually impossible to say on which pair of autosomes the trait is located. Even more rarely can we determine exactly where on a particular autosome the gene for a given trait is.

The mapping of human autosomes is indeed a formidable undertaking, even with the assistance of some recently developed experimental techniques that have made the task somewhat more approachable. Until recently the construction of human chromosome maps depended first on the demonstration that two traits were linked and therefore on the same chromosome. This is not easily demonstrated in human pedigrees because not often are the data concerning two traits clear-cut enough and common enough to allow determination of whether the combinations observed in a given lineage are best explained by assuming that the traits are linked and that they undergo recombination by crossing over a certain percentage of the time. But even if linkage can be established this tells nothing about the particular autosome on which the genes that determine the traits are located. That can be determined if the frequencies of linked genes correspond with the presence of a particular autosomal abnormality but they very rarely do.

More recently, the technique of *somatic cell hybridization* has been applied to the mapping of human chromosomes. This technique is the experimental fusion of body cells (not sex cells) from other mammals, especially mice, with human cells (Figure 1-17). As the hybrid cells divide, the human chromosomes are selectively and progressively lost, and *clones* of cells, which contain a full complement of mouse chromosomes and a few or a single human chromosome, are produced. Under such circumstances it is sometimes possible to detect the presence of enzymes known to be produced by humans but not by mice. One can then conclude that the gene responsible for the enzyme is located on the particular human chromosome present in the hybrid cell. And, as shown in Figure 1-18, recently developed staining techniques make it possible to distinguish individual autosomes even if they are isolated from one another. By using these techniques is has been possible to assign at least 50 human genes to 18 different autosomes.

Why go to all the trouble of mapping human chromosomes? First of all, detailed maps would be of enormous use in genetic counseling and in prenatal diagnosis. In fact, what meager knowlege we have of the map has already been applied in identifying the presence of hard-to-detect detrimental traits in unborn fetuses. For example, the gene for myotonic dystrophy (characterized by wasting of muscles and by other abnormalities) is closely linked to an easily detected "marker trait" that can be detected in fetal cells obtained by amniocentesis (see Chapter 5). A second reason for mapping chromosomes is

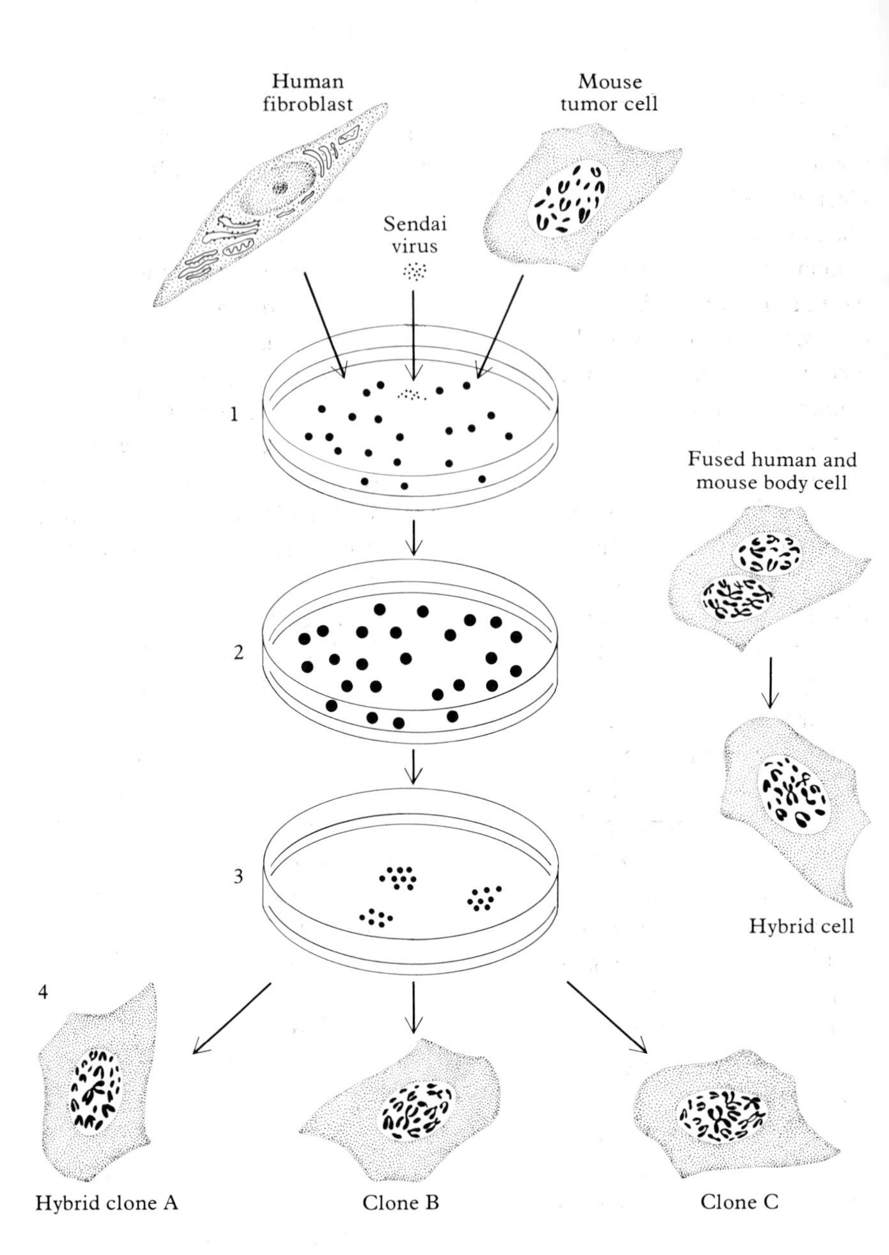

1–17
The techniques of somatic cell hydridization. Human fibroblasts are mixed with mouse tumor cells and incubated with various chemicals and a virus that enhances the fusion of the cells. Hybrid cells result and they can be separated into clones, each of which has special properties, depending on which chromosomes it contains. (From "Hybrid cells and Human Genes," by Frank H. Ruddle and Raju S. Kucherlapati. Copyright © 1974 by Scientific American, Inc. All rights reserved.)

that the attempt to do so has unexpectedly provided information both on the regulation of genes and on their expression in cells. And third, as a leading investigator has put it, the mapping of human chromosomes is rather like climbing Mount Everest. It is exciting because it is there to be accomplished.

Before we leave the subject of autosomes and turn our attention to sex determination and sex chromosomes, we should first consider some human disorders that result from abnormalities in the total number of autosomes within the nuclei of an individual's somatic cells. The most important of these disorders is the presence of an extra copy of chromosome-21 in addition to the usual pair. This condition is known as *trisomy-21,* or *Down's syndrome.*

Down's Syndrome and Other Abnormalities in the Number of Autosomes

People who have Down's syndrome have surely existed since ancient times, but the condition was not described in detail until 1866. Those who are affected with

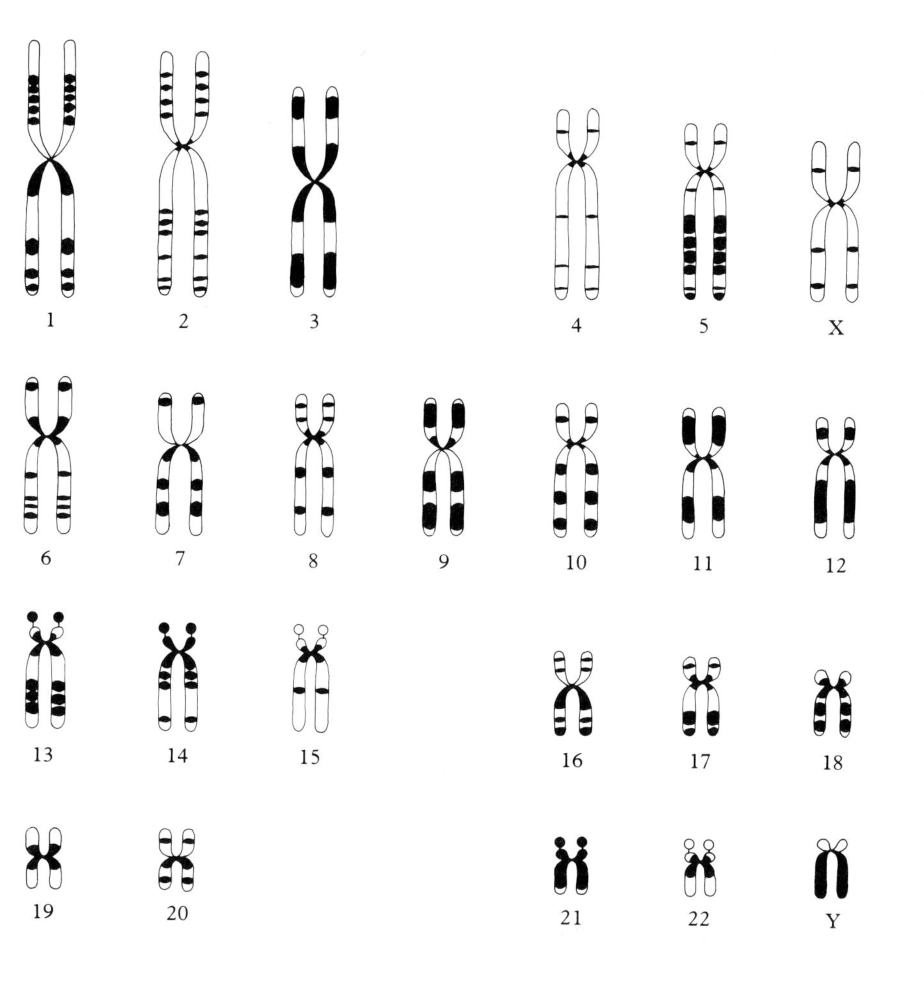

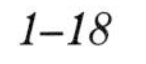

1–18
A simplified diagram of the banding patterns observed among specially stained human chromosomes. (After Drets and Shaw, Proceedings of the National Academy of Sciences, 68, *1971.)*

Down's syndrome are short in stature, frequently have serious malformations of the heart, and usually have characteristically shaped heads with distinctive eyelids and faces. More important, people who have Down's syndrome are almost without exception severely mentally retarded. (To the Europeans who first described the condition, the characteristic appearance of the eyelids suggested the facial features of Mongoloid peoples, and the condition was referred to as "mongolism" or "mongoloid idiocy." In fact, the eyelids of people who have Down's syndrome are quite different from those of people belonging to Mongoloid races, and the earlier, inaccurate terminology has therefore been dropped.)

Down's syndrome occurs sporadically. That is, affected individuals are usually the offspring of normal parents. After it was known that the incidence of the disease varies directly with the age of the mother at the time of birth (as discussed later, older mothers have a greater tendency to produce affected offspring), it was assumed that the syndrome was produced by an unfavorable interaction between mother and fetus during the course of pregnancy. Then, in 1959 (shortly after techniques for seeing human chromosomes had been perfected) some French investigators discovered that the somatic cells of those who had Down's syndrome contained 47 chromosomes, one more than usual. As shown in Figure 1-19, the extra chromosome is a rather small autosome belonging to "Group G," and by convention it is designated chromosome-21.

How do the somatic cells of people who have Down's syndrome end up with an extra chromosome-21? In brief, trisomy-21 is usually the result of an accident that occurs either during meiosis or during the first few cell divisions that take

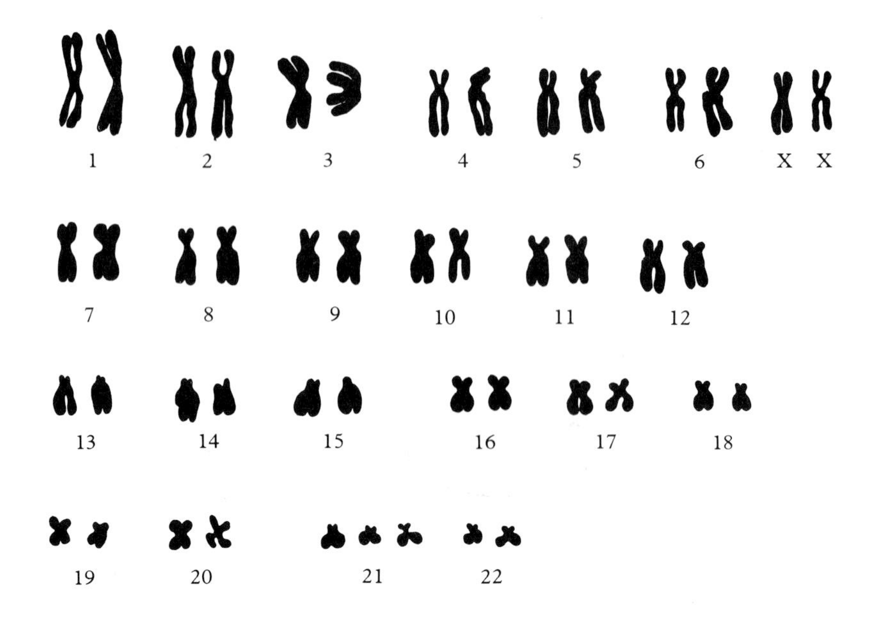

1–19
Down's syndrome most often results because of the presence of an extra copy of chromosome-21, as in the karyotype shown here.

place after fertilization. What happens is this: either duplicated pairs of chromosomes fail to separate during the first division of meiosis or duplicated chromosomes fail to sort out equally in the second division. Either way some sex cells receive two chromosomes-21 and some receive none. This failure of chromosomes to sort out properly during cell division is called *nondisjunction* (see Figure 1-20).

What happens if nondisjunction occurs during the production of a human egg that is then fertilized by a normal sperm? With regard to Down's syndrome, there are two possibilities. If the abnormal egg contains two chromosomes-21 to

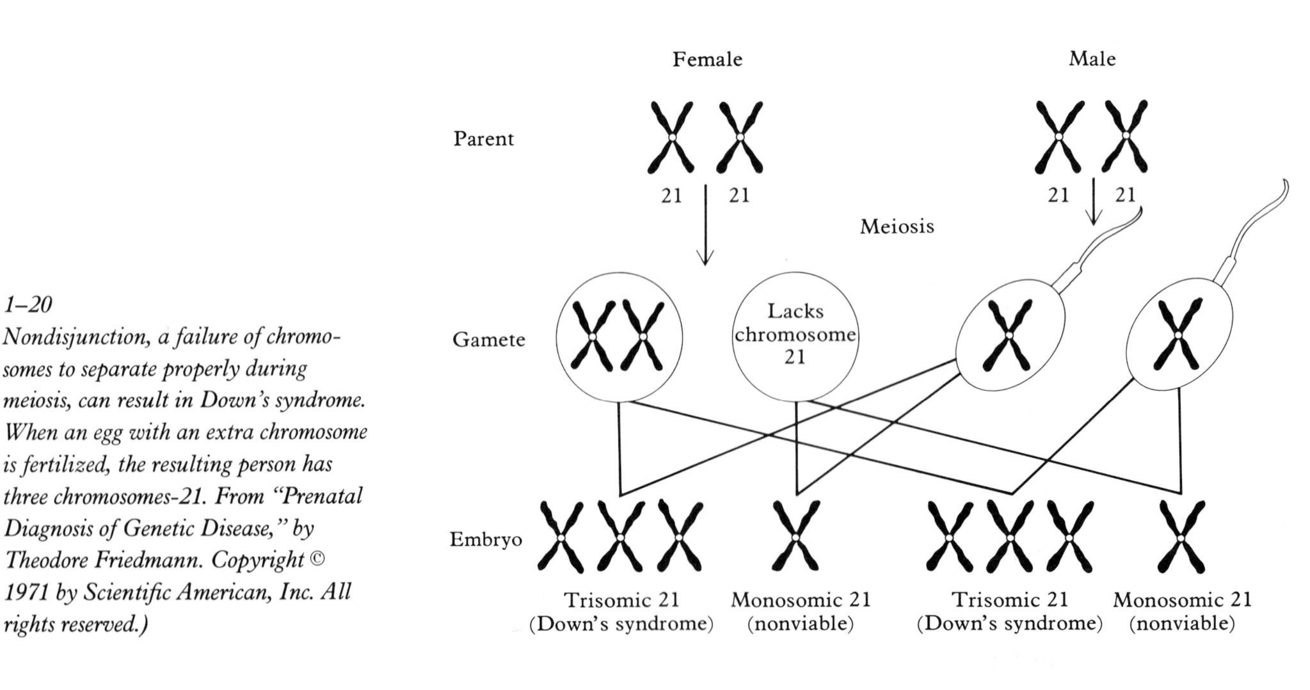

1–20
Nondisjunction, a failure of chromosomes to separate properly during meiosis, can result in Down's syndrome. When an egg with an extra chromosome is fertilized, the resulting person has three chromosomes-21. From "Prenatal Diagnosis of Genetic Disease," by Theodore Friedmann. Copyright © 1971 by Scientific American, Inc. All rights reserved.)

begin with, then on fertilization it becomes trisomic, and if the pregnancy goes to completion a child with Down's syndrome will result. (As discussed in Chapter 5, it is now possible to detect the presence of Down's syndrome while a fetus is only a few months old.) If the egg had no chromosome-21 to begin with it has only one copy after fertilization; this apparently leads to early spontaneous abortion of the developing fetus because no living person who has a single copy of chromosome-21 has been reported. The result is the same if a normal egg is fertilized by an abnormal sperm produced by nondisjunction in the male. Also, nondisjunction can occur in the first few cell divisions following fertilization of a normal egg by a normal sperm. In this case, the cells with a single chromosome-21 die off, and development of the embryo proceeds by further division of those cells trisomic for chromosome-21.

As we shall soon see, variations in the number of human sex chromosomes generally produce abnormalities that are less severe than those produced by the presence of abnormal numbers of autosomes. Trisomy for chromosomes 13 and 18 is known to occur, but those in whom it occurs usually die as infants. Trisomy for other autosomes is apparently lethal before birth because it is not observed among living human beings. Also, deletions of entire autosomes are very rarely seen. In fact, the deletion of even part of an autosome often produces serious, even fatal, abnormalities. A large proportion of the 10 to 15 percent of all pregnancies that terminate in spontaneous abortions before 20 weeks of gestation do so because of chromosomal abnormalities that are incompatible with normal development.

The fact that trisomy-21 is compatible with life no doubt relates at least in part to the small size of the chromosome present in triplicate. Down's syndrome occurs with surprising frequency. It is observed in one out of 500 or 600 births, and has been detected in up to one in 40 fetuses aborted before 20 weeks of gestation. And the chances for the occurrence of trisomy-21 increase with the age of the mother (Figure 1-21). Why is this so?

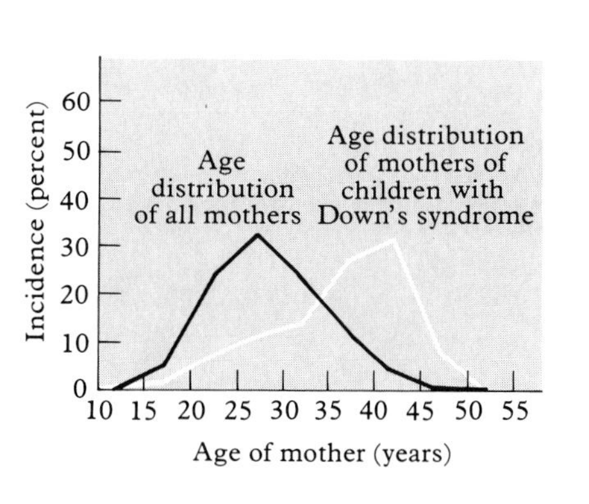

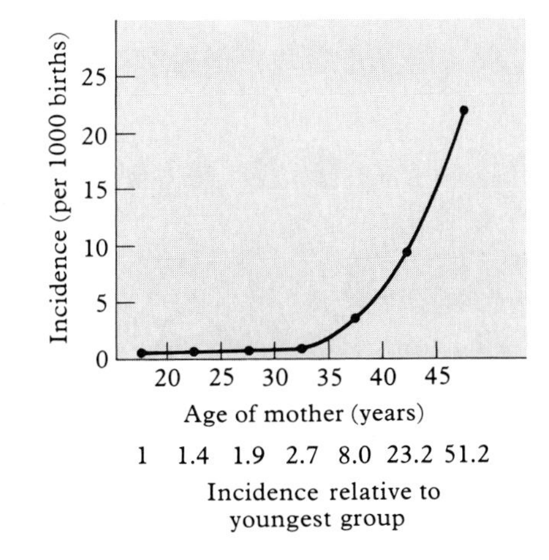

1–21
Left, the age distribution of mothers of children who have Down's syndrome compared to that of all mothers. Right, data about 1,119 cases of Down's syndrome recorded in Victoria, Australia. (Left, after Dr. Victor McKusick; right, after Collman and Stoller, Amer. J. Public Health, 52, *1962.)*

First, it is generally believed that the nondisjunction that leads to trisomy-21 occurs most often in the egg and not in the sperm cell. Perhaps this is because egg cells can sit in the ovary for decades before being ovulated and thus undergo some kind of metabolic or physical damage that later leads to nondisjunction in meiosis. At any rate, the incidence of Down's syndrome increases with maternal age even if the age of the fathers is constant. And, Down's syndrome occurs with a rather characteristic frequency among the offspring of women of the same age, no matter how old their husbands are.

Although maternal age is generally associated with the incidence of Down's syndrome, certain circumstances greatly increase the chance that a particular pair of parents will produce an affected child at any age. These cases of Down's syndrome are not due to trisomy for chromosome-21, but rather are the result of an abnormality of meiosis known as *translocation,* which results in the fusion of two normal chromosomes, or at least parts of chromosomes. Most often the long arms of chromosomes 21 and 14 become fused to form a larger, composite chromosome known as *translocation chromosome.* A person who has one chromosome-21, one chromosome-14, and one translocation chromosome is phenotypically normal even though possessed of only 45 chromosomes (Figure 1-22). This is because all of the genetic material is represented and none is in excess. Such a person is called a carrier of the translocation chromosome. (The loss of the short arms of chromosomes 21 and 14 apparently does not have much effect on a person's phenotype.)

Those who are carriers of a translocation chromosome can produce sex cells of two types—those containing the translocation chromosome and those without

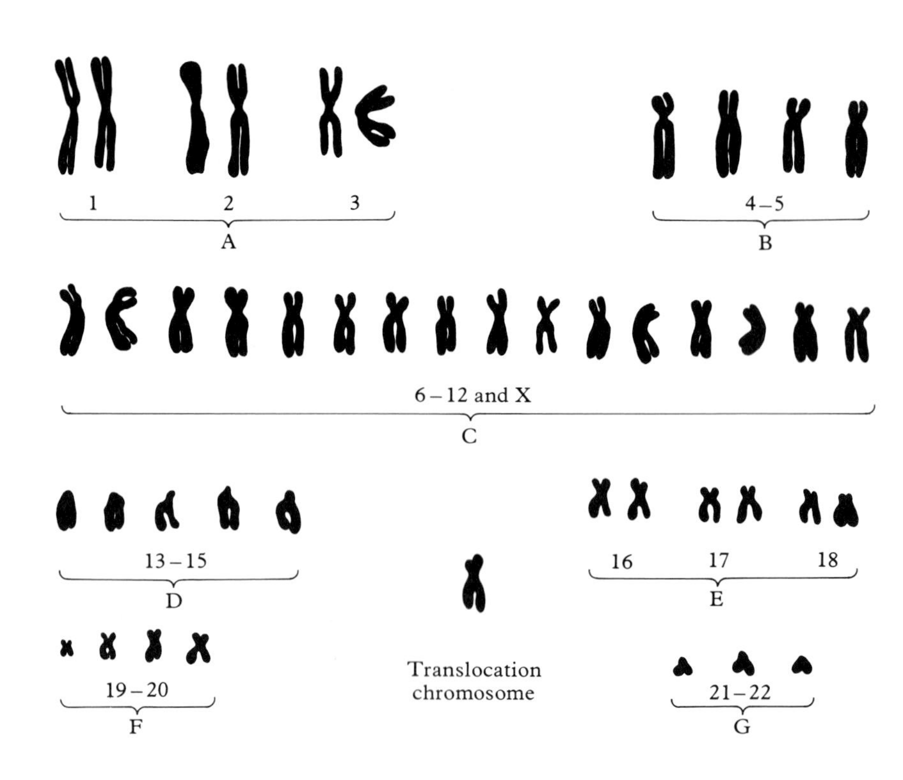

1–22
The 45 chromosomes of a woman who is a translocation carrier of Down's syndrome.

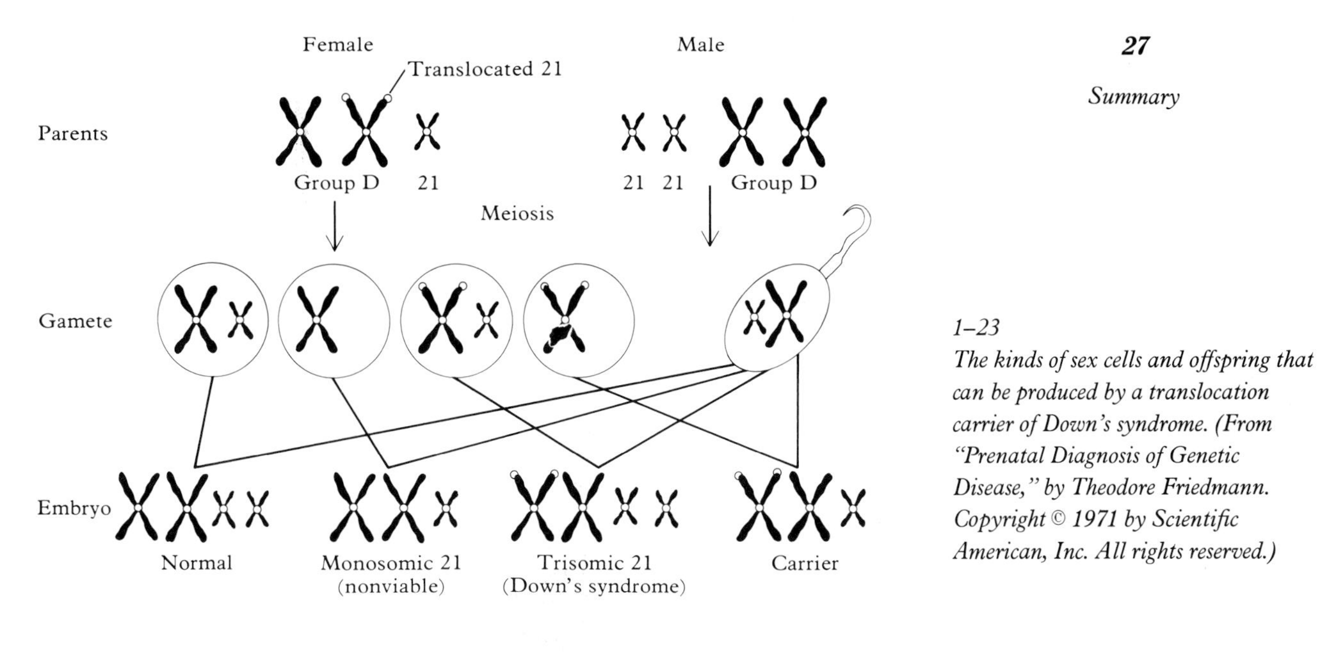

1–23
The kinds of sex cells and offspring that can be produced by a translocation carrier of Down's syndrome. (From "Prenatal Diagnosis of Genetic Disease," by Theodore Friedmann. Copyright © 1971 by Scientific American, Inc. All rights reserved.)

it (Figure 1-23). Either sex cells without the translocation chromosome are normal or they die before or shortly after fertilization because they contain chromosome-21 in but a single copy. On the other hand, after they are fertilized, sex cells containing the translocation chromosome result in the chromosome combinations shown in Figure 1-23. Overall we would expect Down's syndrome to occur regularly in the offspring of people who are translocation carriers, and this is what happens. For this reason, if a child with Down's syndrome is born to relatively young parents, they are advised to have their chromosomes studied to determine whether either parent is a translocation carrier for Down's syndrome and therefore likely to produce another affected child.

With this discussion of autosomes and their abnormalities as background, we are now prepared to take up the subjects of sex determination and sex chromosomes. We do so in the following chapter.

Summary

Species retain their identity in nature because they are isolated from one another behaviorally and genetically. Variation between and within species is due to differences in genetic programs, and variation within a species is not randomly distributed. Some traits show definite patterns in their inheritance, as Mendel discovered.

Mendel proposed that alternate traits were passed from parents to offspring as independent units that are now called genes. He found that the factors responsible for some traits are dominant whereas others are recessive and that he could predict what would happen in crosses involving several traits.

The discovery of the behavior of chromosomes during cell division provided a physical basis for Mendel's patterns. Human beings usually have 46 chromosomes in 23 pairs. Of these, 22 pairs are autosomes and the remaining pair are sex chromosomes.

In human pedigrees, autosomal dominant and recessive traits are usually associated with diseases or abnormalities. Dominant traits vary more in expressivity than recessive ones. Recessive traits can be manifested in a lineage by means of consanguineous marriages. Many human traits are determined by more than one pair of genes, and the expression of most traits depends in part on an individual's environment.

Human chromosomes can be mapped by determining the frequency with which pairs of genes undergo recombination during meiosis or by experiments on somatic cell hybridization. Meiosis results in sex cells that contain one half of a complete set of chromosomes, and recombination helps to provide the raw material for evolution. The mapping of human chromosomes has barely begun, but our knowledge is increasing rapidly.

Abnormalities in the number of autosomes usually produce severe phenotypic abnormalities. The most common of these is Down's syndrome, which occurs more frequently among the offspring of older mothers and which can result from either nondisjunction or translocation.

Suggested Readings

The first five references are for those who would like to learn more about genetics in general. Each of these books is highly recommended. Some are more difficult than others, but all should be understandable to those who can make it through the present volume.

1. *Principles of Human Genetics,* 3d Ed., by Curt Stern. W. H. Freeman, 1973.
2. *Human Genetics,* 2d Ed., by Victor A. McKusick. Prentice-Hall, 1969.
3. *Genetics, Evolution, and Man,* by W. F. Bodmer and L. L. Cavalli-Sforza. W. H. Freeman, 1976.
4. *Heredity Evolution and Society,* by I. Michael Lerner and William J. Libby. W. H. Freeman, 1976.
5. *An Introduction to Genetic Analysis,* by David T. Suzuki and Anthony T. F. Griffiths. W. H. Freeman, 1976.
6. "Chromosomes and Disease," by A. G. Bearn and James L. German, III. *Scientific American,* Nov. 1961, Offprint 150. How advances in the visualization of human chromosomes opened up a new frontier in the study of human heredity.

7. "The Mapping of Human Chromosomes," by Victor A. McKusick. *Scientific American,* April 1971. How human chromosomes can be mapped by applying statistical techniques to data from pedigrees.
8. "Hybrid Cells and Human Genes," by Frank H. Ruddle and Raju S. Kucherlapati. *Scientific American,* July 1974, Offprint 1300. How the fusion of human somatic cells with the cells of other mammals can yield information that can be used in mapping human chromosomes.

Seventeen of the people in this 1894 photo are descendants of Queen Victoria (seated, center). She and the other two women indicated by an asterisk were carriers of hemophilia, a disease that is inherited as an X-linked recessive trait and that is characterized by a prolonged blood clotting time. The other two women are Princess Irene (Henry) of Prussia (right), and Princess Alix (Alexandra) of Hesse (left) who later married Nicholas II, the last tsar of Russia. (The future tsar is standing beside Alexandra.) A pedigree of hemophilia in royal families of Europe is found later in this chapter. (Courtesy of the Gernsheim Collection, Humanities Research Center, University of Texas, Austin.)

CHAPTER 2

Sex Determination and Sex-Linked Traits

In the shallow offshore waters of several continents live some rather drab-looking but remarkable snails known as slipper shells (Genus *Crepidula*). What makes these animals remarkable is the way in which their sex is determined. To simplify a little, the sex of a slipper shell depends on where it happens to land when it settles down to become an adult.

All young slipper shells are males, and they propel themselves through the water by means of winglike structures, as shown in Figure 2-1. Nonetheless, when they lose their wings and settle to the bottom to begin their adult lives, the young males are transformed into females. That is, the young males are transformed into females if they do not land on a female. Young male slipper shells that land on females remain males throughout their lives, unless they become detached. If a mature male slipper shell becomes detached from a female, the male automatically changes into a female, as it would have done had it not landed on a female in the first place.

What does all of this accomplish for the slipper shell? In brief, this unusual method of sex determination makes it likely that males and females will live in the same area and that they will therefore mate with one another.

The slipper shell's method of sex determination is intriguing because it is so unusual. The sex of most animals, especially those with which we are more familiar, is determined at the time of fertilization and remains the same

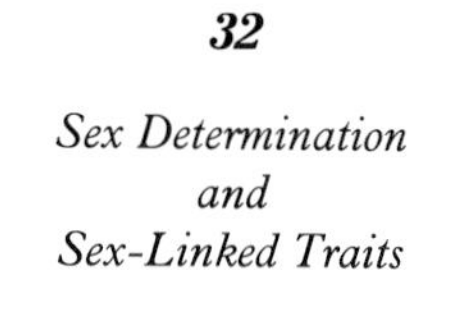

throughout life. Unlike slipper shells, most animals cannot adjust their sex according to circumstances. Rather, sexual reproduction among animals that have separate sexes usually depends on individuals of opposite sex finding and mating with one another. And under such circumstances it is usually advantageous for a species to have about as many males as females.

Most of us would probably agree that the human species has roughly equal numbers of males and females, as do most other species that reproduce sexually. But the biological mechanism underlying the maintenance of this familiar, nearly equal distribution of the sexes eluded biologists until early in this century. At that time, investigators first turned their attention to male–female differences in chromosomes.

In 1902 it was discovered that the body cells of female grasshoppers contain one more chromosome than those of males. Shortly thereafter, the female's extra chromosome was rather romantically dubbed the *X chromosome* (X for unknown) and it was suggested that the presence or absence of the X chromosome determined whether a grasshopper was a female or a male. Since then it has been learned that chromosomes do have a role in sex determination for the great majority of living things that have separate sexes, including most animals. But it turns out that male and female grasshoppers (and some of their close relatives)—with their unequal numbers of chromosomes—are the exceptions rather than the rule.

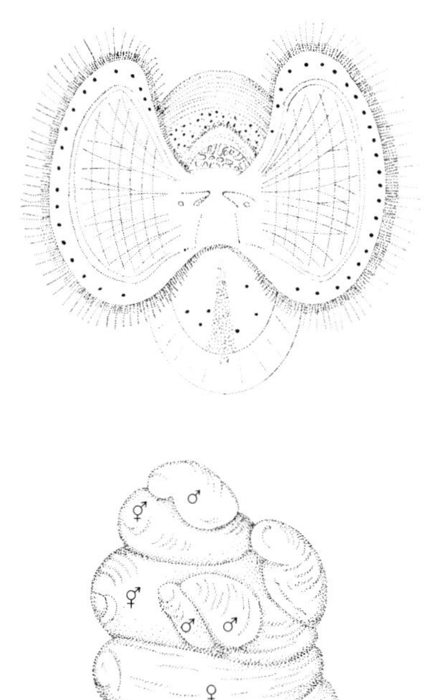

2–1
Top, an immature male slipper shell that has not yet shed his wings, as seen from below. Bottom, a cluster of slipper shells (♀ = female, ♂ = male, ⚥ = individual of intermediate sex).

Male and female animals of the same species generally have the same number of chromosome pairs. But although all of the pairs are matched in females, males have one pair that does not match. As discussed in Chapter 1, the chromosome pairs that match in both sexes are called autosomes, and the members of the remaining pair, which match in females but not in males, are called *sex chromosomes.* For most animals, including all mammals, females are said to be of sex chromosome constitution *XX* and males of *XY.* (In birds the sex chromosomes match in males but not females.)

Figure 2-2 compares the chromosomes of normal males and females. Notice that the X and Y chromosomes are easily distinguished from one another, as, generally, are men and women. In this chapter we discuss the chromosomal basis of maleness and femaleness in human beings and then consider the patterns of inheritance and special properties of traits determined by genes located on sex chromosomes. We begin by discussing the observed human sex ratio and how it relates to the XY mechanism of sex determination.

The Human Sex Ratio

The XY chromosome mechanism of sex determination yields a reliable and nearly equal distribution of the sexes because of the sorting out of chromosomes during meiosis. You will recall that the body cells of women contain 22 pairs of autosomes and two X chromosomes. So when meiosis reduces the number by half during the production of sex cells, eggs that contain 22 unpaired autosomes and a single X chromosome are produced. Thus all normal eggs have an X chromosome. But not so sperm. The body cells of males contain 22 pairs of autosomes, one X chromosome, and one Y chromosome. So when men produce sex cells, meiosis results in two kinds of sperm with regard to sex chromosomes. All human sperm normally contain 22 unpaired autosomes, and on the average

2–2
The chromosomes of normal males and females. Left, males have 22 pairs of autosomes, one X chromosome, and one Y chromosome. Right, females have 22 pairs of autosomes, and two X chromosomes.

half of the sperm have an X chromosome and half have a Y chromosome. Thus, if X-bearing and Y-bearing sperm fertilize normal eggs and result in normal development about equally often, then we would expect that the sexes ought to be about equally distributed, as shown in Figure 2-3. But are they?

Relevant data concerning the number of males and females born in recent decades throughout the world exist and they reveal some rather surprising facts. By convention, the sex ratio is usually reported as the number of males per one hundred females. Among Caucasians in the United States the data show that approximately 106 boys are born for every 100 girls, so the sex ratio at birth is 106. The sex ratio at birth varies somewhat from country to country and it can vary from one racial group to another. For example, the ratio is 113 in Korea, whereas it is about 102.6 among American blacks. Nonetheless, the worldwide data show that on the average more males than females are born in every time interval for which reliable data exist.

The sex ratio at the time of fertilization, known as the *primary sex ratio,* is not necessarily the same as the ratio at the time of birth, the *secondary sex ratio.* Human males have at least a slight numerical edge on females at the time of birth. Is this because more females than males die as embryos? Apparently not, for statistical studies on unborn fetuses have revealed that the primary sex ratio is even higher than the secondary—perhaps as high as 130. Of course, there are many sources of error in determining the sex of unborn fetuses, and the task of assigning a definite sex to the youngest embryos is most difficult. The possibility remains that the observed primary sex ratio results from a large number of female deaths at *very* early stages of development, a period for which we have very little data. But overall, it appears likely that human males do have a numerical advantage over human females both at the time of fertilization and at the time of birth. How can this rather unexpected observation be explained?

If there really are more males than females at the time of fertilization, then several explanations are possible. For example, it may be that Y-bearing sperm win over X-bearing sperm in the race to the waiting egg. It has been suggested

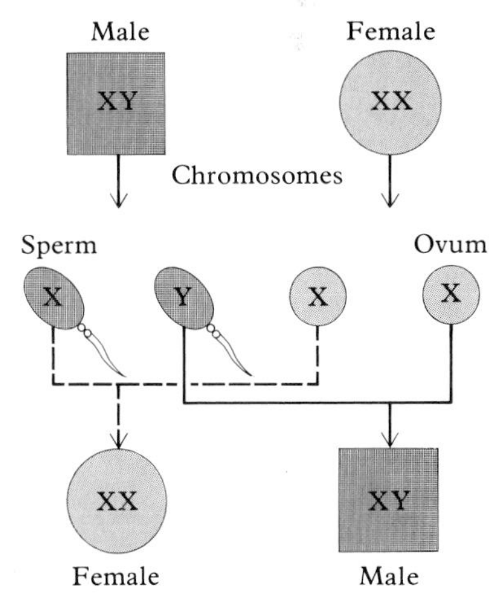

2–3
The chromosomal basis for the existence of nearly equal number of men and women. (From "Sex Differences in Cells" by Ursula Mittwoch. Copyright © 1963 by Scientific American, Inc. All rights reserved.)

that because the Y chromosome is smaller than the X, Y-bearing sperm are lighter and therefore able to swim faster than X-bearing sperm. But, swimming is not the primary means by which human sperm reach the egg. Rather, muscular contractions and ciliary currents within the female reproductive tract are primarily responsible for transporting human sperm to the oviduct (Fallopian tube), where fertilization occurs. Perhaps Y-bearing sperm really do have some as yet unidentified advantage over X-bearing sperm that accounts for the greater numbers of male conceptions and births. (The two types of sperm are produced in approximately equal numbers.) One explanation for the preponderance of male conceptions could be that the physiological environment of the female reproductive tract favors the survival of Y-bearing over X-bearing sperm. Still another possibility is that the surface of the egg attracts Y- more strongly than X-bearing sperm, and this by no means exhausts the possible explanations. In the end, we really do not know what accounts for the greater numbers of males at conception and birth; all we know is that they exist.

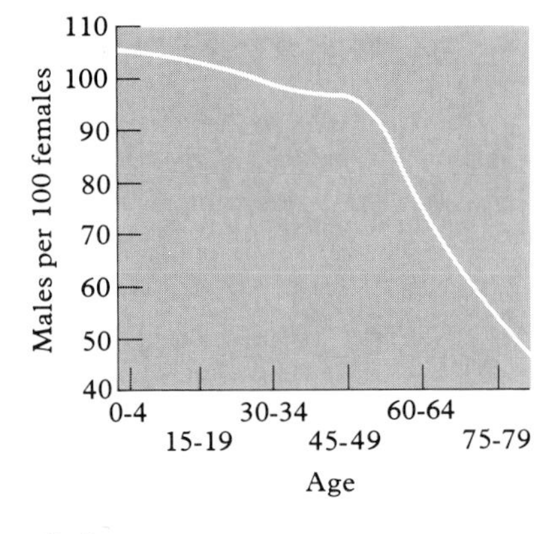

2–4
Data concerning the secondary sex ratio in England and Wales in 1960. (From A. S. Parkes.)

Human males start life with a numerical advantage over females, but they finish a weak second. This is because more males than females die at every stage of life from conception to old age. Thus the numerical advantage of males at birth becomes progressively smaller until at a certain age the sexes exist in equal numbers. The exact age at which this occurs varies from population to population. As shown in Figure 2-4, in some countries males and females are in equal numbers by the time they reach 30 years of age; in others this happens as early as age 18 or as late as age 55. At whatever age it occurs, the equal number of males and females is not maintained. Females soon become the clear numerical majority.

It has been suggested that the male's greater susceptibility to death at every age may be related to the fact that men have only one X chromosome. (As we shall soon discuss, the Y chromosome is very nearly a genetic blank for inherited human characteristics.) It is argued that males, with their single X chromosome, are more vulnerable to the effects of deleterious genes located on it. Even if they do have a deleterious gene on one X chromosome, women are likely to have a corresponding normal allele on their other X chromosome and thus are less likely to be severely affected. Once again, we do not know whether the human male's possession of a single X chromosome is really related to his greater constitutional weakness at every stage. But arguments from population genetics and recent advances in our understanding of the genetics of sex chromosomes have made it seem unlikely that men are at much of a disadvantage simply because of possessing a single X chromosome. As we shall mention later in this chapter, in normal women only one X chromosome is genetically active in each cell; the other X chromosome is nonfunctional.

Voluntary, predictable changes in the primary and secondary sex ratios may soon be a reality. Technological means of altering the sex ratio of the human population already exist. For example, the secondary ratio could be altered by selective early abortion of fetuses of unwanted sex, but this means of controlling the sex of offspring is not likely to become widely accepted or widely available. A more appealing approach to voluntary sexual preselection of offspring is separation of X- and Y-bearing sperm outside the body followed by artificial insemination of either X- or Y-bearing sperm. This technique has been successfully employed in animal husbandry and has been made more reliable by the recent development of a staining method that can identify the Y chromosome in living human cells, including sperm. Other less complicated techniques, such as the use of chemicals or prophylactics designed to block the entry of X- or Y-bearing sperm into the female reproductive tract, would probably be more widely accepted, but they are only theoretical possibilities at the present time.

What would happen to the sex ratio at birth if sexual selection were freely available in the United States? Of course, we can only speculate, but there is good reason to believe that the present ratio of about 106 male to 100 female births would probably not change much. Based on the data from sociological surveys it has been predicted that what would change would be the probability that the first born child would be a male and the second a female. What effect, if any, this would have on our society remains to be seen, as do any other long-term effects of sexual preselection.

We now return our attention to a discussion of the sex chromosomes themselves and in particular to the relationship between femaleness and the X chromosome and maleness and the Y chromosome.

The Roles of the X and Y Chromosomes in Sex Determination—Abnormalities in the Number of Sex Chromosomes

What is the critical genetic difference between normal men and women? From our discussion so far, the answer would appear to be: women are of sex chromosome constitution XX and men are XY. But this tells us nothing of the exact relationship between the X and Y chromosomes and human maleness and femaleness. After all, the sexes are distinguished chromosomally not only by the male's having a Y chromosome, but also by his having only one X chromosome instead of two. Is it the presence of two X chromosomes that determines femaleness and the presence of a single X that results in maleness? Or does the Y chromosome determine maleness? These questions can now be answered. As so often happens in biology, our understanding of the normal process came about through the study of those persons in whom the normal mechanism of chromosomal sex determination had gone awry.

In 1949 it was discovered that two well-known but puzzling human afflictions, *Turner's syndrome* and *Klinefelter's syndrome,* are the result of abnormal sex-chromosome constitutions. (A syndrome is a group of signs and symptoms that occur together and characterize a particular abnormality.) Those who have Turner's syndrome are phenotypic females, which is to say that their genitalia are recognizably female. But these women are generally sterile because most have underdeveloped uteruses and no functional ovarian tissue. Other features of Turner's syndrome are short stature, rather distinctive facial features, a broad shieldlike chest, and a peculiar webbing of the neck. Those who are afflicted with Turner's syndrome have a single, unpaired X chromosome and are thus of sex chromosome constitution XO.

This discovery was made still more interesting when in the same year Klinefelter's syndrome was found to be associated with the sex chromosome constitution XXY. Those who have Klinefelter's syndrome are phenotypic males, but they usually have very small testes and are sterile. Most are also very long-legged and have breast development resembling that of a female.

How does an individual come to have an XO or XXY sex chromosome constitution? Usually by nondisjunction, a term you will recall from our discussion of Down's syndrome in the preceding chapter. Either the sex

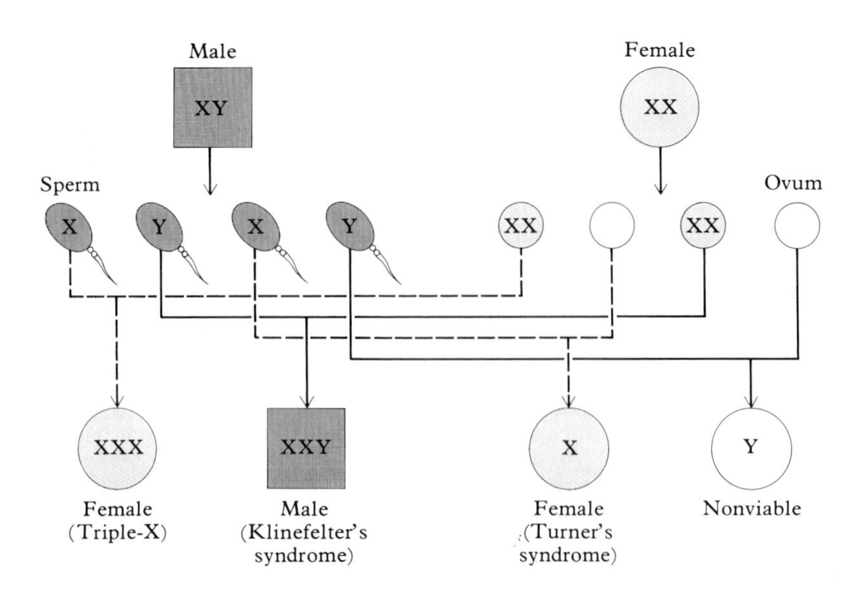

2–5
How nondisjunction can result in certain abnormalities in the number of sex chromosomes. (From "Sex Differences in Cells," by Ursula Mittwoch. Copyright © 1963 by Scientific American, Inc. All rights reserved.)

chromosomes fail to sort out properly during the production of sex cells by meiosis, or they fail to sort out properly in the first few cell divisions following fertilization (see Figure 2-5).

The discovery of the chromosomal basis of Turner's and Klinefelter's syndromes was exciting, not only because the existence of the abnormal sex chromosomes could be explained as the results of nondisjunction, but also because of the insights it provided into the workings of the normal XY mechanism. That those who have XO chromosomes are phenotypic females, whereas people who have a sex chromosome constitution of XXY are phenotypic males suggests that the Y chromosome determines maleness and that the reason those who have XO chromosomes appear female is that in the absence of the Y chromosome even one X is enough to result in a female phenotype.

Klinefelter's syndrome occurs once in every 400 to 600 male births, whereas Turner's syndrome occurs once in about every 3,500 female births. (But Turner's syndrome is observed in many fetuses that undergo spontaneous abortion before twenty weeks of gestation.) In the years since 1949 a rather startling number of cases of human abnormalities caused by unusual numbers of sex chromosomes have been reported. Most of these conditions occur less frequently than Turner's and Klinefelter's syndromes. Many of those who are affected are mentally retarded, and most have physical abnormalities. Sex chromosome constitutions of XXX, XXXX, and XXXXX have been reported, and those who have them are all phenotypic females. On the other hand, sex chromosome constitutions of XXXY, XXXXY, and XXXXXY are also known (Table 2-1) and all belong to phenotypic males. Thus it seems well established that in human beings the Y chromosome determines maleness and its absence results in a phenotypic female, as long as at least one and up to five X chromosomes are present.

Further evidence of the male-determining role of the human Y chromosome comes from the study of persons known as *genetic mosaics.* These persons are remarkable in that their bodies consist of two or more cell lines (that is, cells with different numbers of chromosomes) side by side. Mosaics are the result of accidents that occur during cell division. Most often the accident is either nondisjunction or the accidental loss of a particular chromosome, and the error usually occurs in the first few cell divisions following fertilization. Mosaics *can*

TABLE 2–1

How sex chromosomes relate to sex phenotype. (Barr bodies are discussed later in this chapter.)

SEX CHROMOSOME CONSTITUTION	SEX PHENOTYPE	NUMBER OF BARR BODIES
XX (normal woman)	Female	1
XY (normal man)	Male	0
XO (Turner's syndrome)	Female	0
XXY (Klinefelter's syndrome)	Male	1
XYY (see chapter 20)	Male	0
XXX	Female	2
XXXY	Male	2
XXXX	Female	3
XXXXY	Male	3
XXXXX	Female	4
XXXXXY	Male	4

result from double fertilizations or from the fusion of two embryos very early in development, but this rarely happens.

Mosaics for sex chromosomes are encountered more frequently than those for autosomes (although the latter do occur). In fact, the first person with Klinefelter's syndrome whose chromosomes were studied was a sexual mosaic. Some cells in his bone marrow were XXY, whereas others were XX. This sex chromosome constitution is thus designated XX/XXY.

More recent studies have shown that mosaics in whom only X chromosomes are different (for example, XO/XX, XX/XXX, and XXX/XXXX, all of which have been reported) are phenotypic females (Table 2-2). Sexual mosaics in whom every cell has a Y chromosome, including XY/XXY, XY/XXXY, and XXXY/XXXXY, are, as you might expect, phenotypic males. (Nonmosaic persons of genotype XYY are also phenotypic males. The XYY genotype is further discussed in Chapter 5.)

Of special interest are sexual mosaics whose bodies are made up of cells of different sex. As summarized in Table 2-2, such individuals are usually phenotypic males if one Y chromosome is present. Nonetheless, some of those who have both male and female cell lines exhibit some of the secondary sexual

TABLE 2–2

Human sex-chromosomal mosaics. The mosaics may combine two or three chromosomal constitutions. Phenotypically the mosaics may be female, male, or mixed.

FEMALE	MALE	MIXED
XO/XX	XY/XXY	XO/XY
XO/XXX	XY/XXXY	XO/XYY
XX/XXX	XXXY/XXXXY	XO/XXY
XXX/XXXX	XY/XXY/XXYY	XX/XY
XO/XX/XXX	XXXY/XXXXY/XXXXXY	XX/XXY
XX/XXX/XXXX		XX/XXYY
		XO/XX/XY
		XO/XY/XXY
		XX/XXY/XXYYY

Source: From *The Principles of Human Genetics,* 3d ed., by Curt Stern. W. H. Freeman and Co. Copyright © 1973.

characteristics of the two sexes simultaneously. For example, XX/XXY mosaics may have one male and one female breast, bearded and unbearded facial areas, and, more important, both testicular and ovarian tissue side by side in the same gonad. People who have both kinds of sex tissue generally have a mixture of male and female features in their external genitalia and are known as *hermaphrodites.* (The word is derived from the names of the Greek deities Hermes and Aphrodite.)

Although most hermaphrodites are mosaics for cells of different sex, nonmosaic hermaphrodites of sex chromosome constitution XX (normal female) and XY (normal male) have been reported. As you know, the human Y chromosome is strongly male determining. How then can we account for the presence of male characteristics in these rare XX individuals and female characteristics in these rare XY individuals?

The most widely accepted explanation is this: the mixture of male and female features that exists in exceptional XX and XY hermaphrodites results from the effects of genes that determine sex but that are located on autosomes. (This assumes that the presence of tumors that secrete sex hormones has been excluded. Such tumors may make the secondary sexual characteristics of normal XX females and XY males resemble those of the opposite sex.)

The existence of autosomal genes that influence sex is further supported by well-documented reports of normal males, who have fathered normal-looking sons, but who are nonetheless apparently of sex chromosome constitution XX! Furthermore, it is known that at least one gene, that for testicular feminization, can override the normal XY mechanism of sex determination, because those who are affected by it are phenotypic females, though their sex chromosomes are XY. The available data concerning the transmission of this rare gene do not allow us to decide whether the gene is located on an autosome or on the X chromosome. Nonetheless, the existence of autosomal genes that play a part in sex determination has been demonstrated many times among nonhuman animals whose XY mechanism functions in the same way as that of humans. Overall, there is little doubt that the 22 pairs of human autosomes do indeed carry genes that play a part in determining sex.

Most of the time the sex-determining effects of autosomal genes are balanced by the normal XY mechanism. Thus normal men and women both have genes for maleness and femaleness, both on sex chromosomes and on autosomes. In men, male-determining genes on the Y chromosome and on autosomes outweigh the female-determining effect of genes on the X chromosome and on autosomes. In women, female-determining genes on the X chromosome and on autosomes outweigh autosomal male-determining genes. (The latter are generally much weaker in their male-determining effects than is the presence of a Y chromosome.) But in rare instances the effects of autosomal male-determining or female-determining genes may override the effects of sex-determining genes located on sex chromosomes, thus resulting in persons whose phenotypes and sex chromosomes are at odds with one another.

This concludes our discussion of human sex chromosomes as they relate to sex determination. We now turn our attention to the genetics of genes that do not influence sex but are nonetheless located on the X and Y chromosomes. Such genes are said to be *sex linked* and they have characteristic patterns of inheritance, as we are about to discuss. The X chromosome probably has about as many genes as an autosome of similar length. But the Y chromosome, in spite of its strong male-determining effect is, as far as we know, nearly completely lacking in other genes.

Y-Linked Inheritance

The pattern of inheritance of genes located on the Y chromosome is very simple. Only men have a Y chromosome and all sons but no daughters receive this chromosome from their fathers. So a man who manifests a trait that is determined by a gene on the Y chromosome will pass the trait on to all of his sons and to none of his daughters.

The restriction of a trait to males is not sufficient evidence to prove the existence of a Y-linked gene. This is because some autosomal traits are expressed only in males (or only in females). But such "sex-influenced" autosomal traits can generally be distinguished from those determined by genes on the Y chromosome because traits that depend on autosomal genes are transmitted by both parents, whereas Y-linked traits neither appear in women nor are transmitted by them.

Of the nearly 2000 human traits known to have a genetic basis, only a few are known to be Y linked. The only human trait known to be definitely determined by a gene on the Y chromosome is the presence on the surface of all male cells of a certain protein, called a histocompatibility antigen (H-Y antigen), that is not found on the surface of the cells of females. H-Y antigen was discovered when it was observed that female mice reject skin grafts from males of the same inbred line. There is evidence that the gene for H-Y antigen and the gene that determines the development of testes may be one and the same. At any rate, the presence of H-Y antigen on the cell surface is now considered a reliable criterion of maleness and it can be used to ascertain the sex of infants whose sex is ambiguous at birth.

The only other trait that has stood the test of time as a probable example of Y linkage is a rather unromantic but harmless one known as hairy ear rims. Affected men have long, stiff hairs on the rims of their ears, as shown by the three Muslim brothers from South India in Figure 2-6. The trait also occurs among Caucasians, Australian aborigines, and, more rarely, among Japanese and Nigerian men. Not all instances of hairy ear rims can be attributed to the effects of Y-linked genes. In some instances genes located on autosomes are clearly reponsible for the trait. Nonetheless, in some groups, especially those from India, Y-linkage appears to be established beyond all reasonable doubt.

The human Y chromosome is not unique in its apparent genetic inertness (aside from its role in sex determination). In some insects the Y chromosome is also known to be nearly devoid of genes other than those that determine sex. On the other hand, some fish have numerous Y-linked genes, as do some mice. In all, the Y chromosome remains rather poorly understood, but we can expect our knowledge of it to increase rapidly. Chromosomes are once again the objects of intensive study, just as they were at the turn of the century. As we learn more about the structure of chromosomes, the role of the Y chromosome in sex determination and in Y-linked inheritance will become more clear.

Although the Y chromosome is a near blank for inherited traits, the X chromosome is far from it. We now turn our attention to the distinctive patterns of inheritance of traits determined by genes located on the X chromosome.

X-Linked Inheritance

About 100 abnormal traits are known to be determined by genes located on the X chromosome. (This means, of course, that the corresponding normal traits are

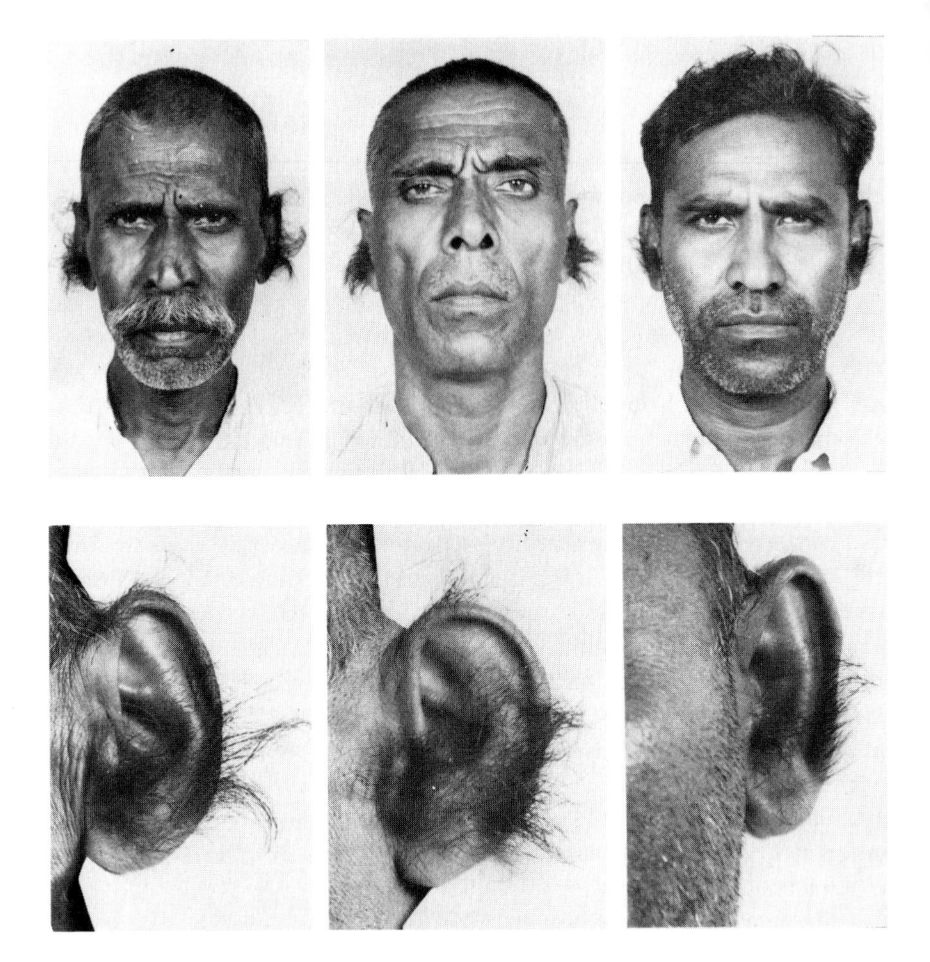

2–6
The strikingly hairy ear rims of three Muslim brothers from South India. This trait is probably determined by a Y-linked gene. (Photo by S. D. Sigamoni, Photography Dept., Christian Medical College Hospital, Vellore. From Stern, Centerwall, and Sarkar, American Journal of Human Genetics, 16. *Copyright © 1964.)*

determined by normal X-linked genes.) The X chromosome is by far the best known of human chromosomes for two reasons. First of all, the patterns of inheritance of X-linked traits are very distinctive, and whenever we observe these patterns in a pedigree we can usually conclude that the gene responsible for the trait is on the X chromosome. (As you know from our discussion of autosomal genes, pedigree analysis usually allows us to decide whether a trait is transmitted as an autosomal dominant or recessive, but tells us nothing about the particular pair of autosomes on which the gene is located.) Second, assigning so many genes to the X chromosome allows us to at least begin to construct a genetic map of this chromosome. As discussed in the preceding chapter, genes located close to each other on a particular chromosome undergo recombination by crossing over during meiosis less frequently than genes that are farther apart. By studying the rates of occurrence of crossovers for two or more X-linked traits we can get an idea of the relative distances between the responsible genes on the X chromosome. Not surprisingly, it is both hard to get and hard to interpret data about recombinations between X-linked genes. Nonetheless, the actual construction of a detailed map of the X chromosome has already begun, as shown in Figure 2-7. This map may seem primitive, but it is by far the most detailed and accurate map of any human chromosome.

Like autosomal traits, those traits determined by X-linked genes may be either dominant or recessive. Nonetheless, X-linked traits have some peculiarities in their patterns of inheritance because of the presence of two X chromosomes in females and only one in males. Females, with their two X chromosomes, may be either heterozygous or homozygous for an abnormal X-linked allele. If the abnormal allele is a dominant one, then a woman heterozygous for it will manifest the trait. But not so if the abnormal allele is recessive. In that case, heterozygous women usually do not manifest the trait, but instead are carriers who appear normal.

On the other hand, males, with their single X chromosome, will always show the effects of an abnormal allele on their X chromosome regardless of whether the trait is inherited as a dominant or a recessive, and in general all males who have abnormal alleles on their X chromosomes manifest X-linked traits about equally. (In other words, males who have X-linked traits are always affected; they cannot be unaffected carriers.)

Of the 100 or so X-linked traits we now know, only a few appear to be inherited as dominants. Included among them are brown discoloration of the teeth, the presence or absence of a particular antigen on the surface of red blood cells, and rickets resistant to vitamin D (a syndrome that is usually characterized by skeletal deformities and by low concentrations of phosphate in the blood).

What are the characteristic features of X-linked dominant inheritance? As shown in Figure 2-8, most women affected by X-linked dominant traits are heterozygous. (Recall that both men and women who manifest autosomal dominant traits are also usually heterozygous.) Women heterozygous for an X-linked dominant trait are affected, and they transmit the trait equally to their sons and daughters. Affected men transmit the trait to all of their daughters (all daughters receive their father's X chromosome), but to none of their sons (no sons receive their X chromosome from their father). In a given pedigree the fact that an affected father does not produce affected sons allows us to distinguish X-linked dominant traits from autosomal dominant ones. In autosomal dominant traits, both affected women and affected men pass the trait on to half of their daughters and to half of their sons. (Work this out for yourself.)

As you may gather from this discussion, the great majority of human abnormalities known to be determined by genes on the X chromosome are inherited as recessive traits. As is also true of X-linked dominant traits, a critical characteristic of X-linked recessive inheritance is the absence of father-to-son transmission. The following are some other characteristics of X-linked recessive inheritance, all of which are summarized in the pedigrees shown in Figure 2-9.

First, virtually all people affected by X-linked recessive traits are men. This is because X-linked traits are rare, and because in order for a woman to be affected, she has to have an abnormal allele on each of her X chromosomes. In

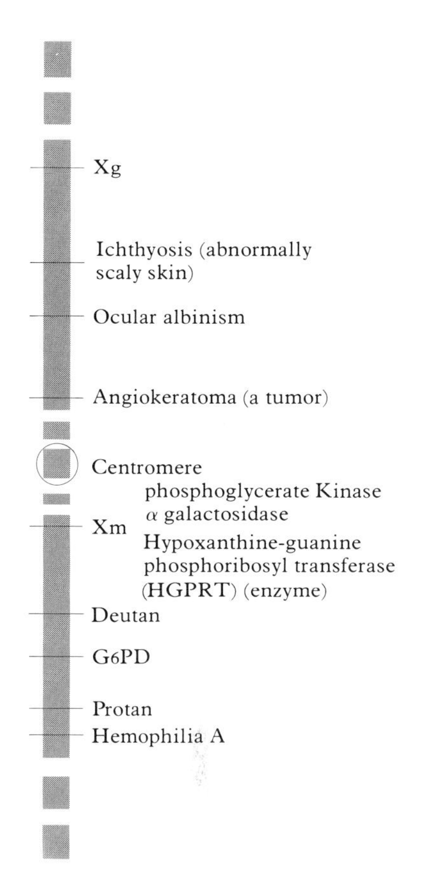

2–7
A tentative map of the human X chromosome. Distances are proportional to the amount of crossing over that takes place between duplicated pairs of X chromosomes during meiosis. Deutan and protan are two forms of color blindness. Xg is a blood group protein, and Xm is a protein found in the blood. There are at least 93 loci known to belong to the X chromosome and almost as many others that are believed to be X-linked. (From Genetics, Evolution, and Man *by W. F. Bodmer and L. L. Cavalli-Sforza. W. H. Freeman and Company. Copyright © 1976.)*

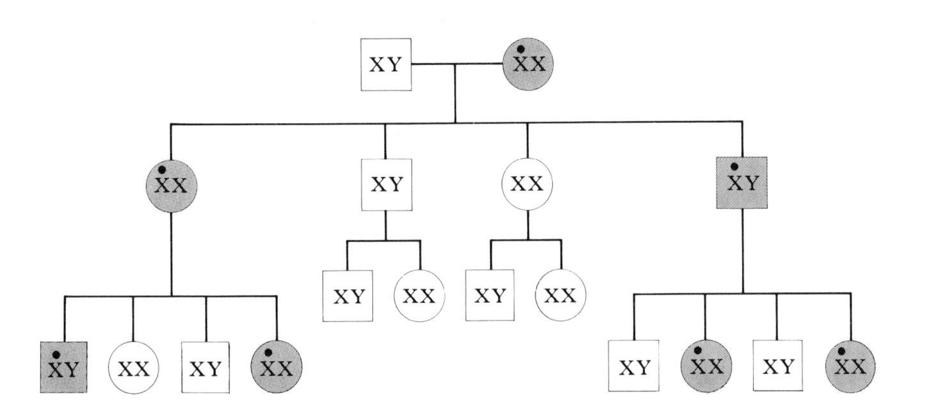

2–8
A pedigree showing X-linked dominant inheritance. Affected individuals are indicated by shaded symbols. The X chromosome bearing the abnormal allele is indicated by a black dot.

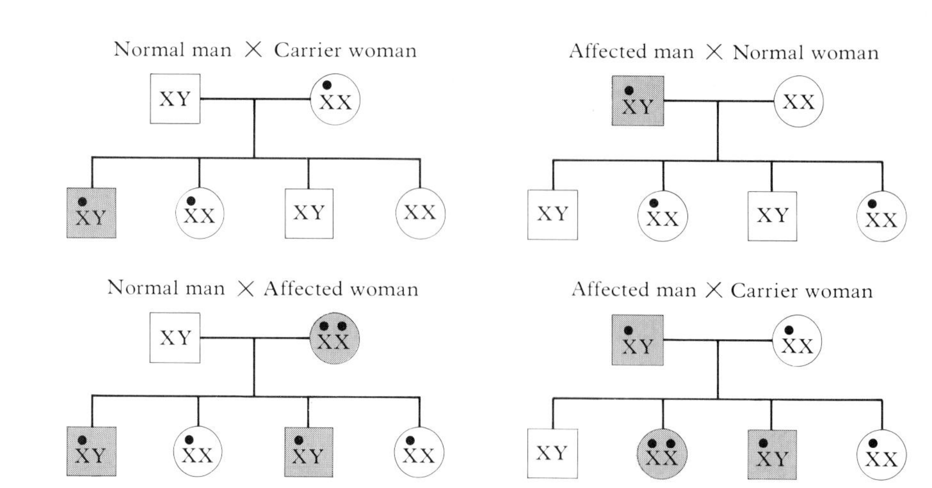

2–9
Pedigrees showing X-linked recessive inheritance. Affected individuals are indicated by shaded symbols. The X chromosome bearing the abnormal allele is indicated by a black dot.

the absence of a spontaneous mutation, this could only occur if her mother were a carrier and her father were affected (Figure 2-9). Such rare, affected women have been recorded in human pedigrees. Second, all of the sons of affected men married to normal women are normal, whereas all of their daughters are carriers. Third, on the average, half of the sons of heterozygous (carrier) women married to normal men are normal, and half are affected (see Figure 2-9).

The most widely publicized of all human pedigrees are probably those in which the X-linked recessive trait known as *classical hemophilia* was transmitted among the royal families of Europe, particularly England. People who have hemophilia are sometimes called "bleeders." Their blood does not clot properly because of a deficiency of one of the many factors that participate in the normal clotting mechanism. (In classical hemophilia the deficiency is in Factor VIII; other types of hemophilia result from deficiencies of different factors.) Those who are affected are usually males (though rare female hemophiliacs have been reported) and they tend to bruise easily and to bleed heavily either into their joints, from their gums, or through open lacerations, often as a result of relatively minor injuries.

The pedigree of Queen Victoria and her descendants is shown in Figure 2-10. Analysis of it reveals that Queen Victoria must have been a carrier for classical hemophilia. One of her sons, Leopold, Duke of Albany, died of hemophilia at age 31. None of Victoria's forebears and neither her husband nor any of her then-living relatives had hemophilia, so the trait probably first appeared as a spontaneous mutation in one of the X chromosomes before Victoria inherited it from one of her parents. Or a mutation could have occurred in one of Victoria's X chromosomes during the queen's early embryonic life. Either way, the pedigree indicates that Victoria must have been a carrier. (The present-day royal family of England is completely free from the gene because the ruling Queen Elizabeth traces her descent through Edward VII, one of Victoria's sons who did not have hemophilia.)

X-linked recessive genes are also responsible for some other rather familiar traits. Of these perhaps the most common are pattern baldness, the common kind of red-green color blindness, and one form of muscular dystrophy, a disease in which the muscles of young males waste away in spite of the presence of an apparently normal nervous system.

For some traits determined by genes located on the X chromosome the distinction between "normal" and "abnormal" is not always clear. One example is the trait known as *G6PD deficiency,* which results from the relative lack of the enzyme glucose-6-phosphate dehydrogenase, an enzyme that participates in carbohydrate metabolism. Affected persons are completely normal under most circumstances, but if they come into contact with certain environmental substances—ranging from the inhalation of the pollen of fava beans to the ingestion of primaquine, a drug used in treating malaria—there may be disastrous results. In the presence of these materials, among others, the red blood cells of affected individuals tend to break open; thus, severe anemia may result.

You will recall that heterozygous carriers of traits determined by genes located on autosomes can usually be identified by some kind of measurement, provided that the underlying biochemical defect is known. It turns out that people who are heterozygous for autosomal traits usually have about half of the normal product of the gene for which they have an abnormal allele. (Thus men and women who are carriers of an allele that results in albinism in homozygotes have about half of the normal concentration of products that have a role in the manufacture of the pigment melanin.) This suggests that autosomal alleles contribute about equally to the total concentration of their normal biochemical product.

This raises an important question. Normal women have two X chromosomes and normal men have only one. Therefore, to return to the example of G6PD, normal women have two normal alleles for the production of the enzyme G6PD, whereas men have only one. Is the concentration of the enzyme in the blood of men therefore only one half of what it is in women? No, it is about the same in both sexes. And this is true not only of the enzyme G6PD, but of the biochemical products of the X-linked genes in general. How can we account for this?

Dosage Compensation in X-Linked Genes—Lyon's Hypothesis

There are at least two possible explanations for the fact that women, with two X chromosomes, have about the same amount of any product determined by an X-linked gene as men, with their single X chromosome. First, the male's single X chromosome could work twice as hard—that is, produce twice the amount of gene product—as each of the female's X chromosomes. (This is true for the fruit fly.) Or second, the activity of one or both of the female's X chromosomes could be less than that of the male's. There is little doubt that for the human species the second explanation is the correct one. Some of the earliest proof for this came from the study of persons who had abnormal sex chromosome constitutions.

In the late 1940s it was discovered that the nondividing cells of female cats contain a small but distinct and stainable blob within their nuclei that is absent from the nuclei of males. This rather mysterious object became known as the *Barr body,* named after one of the first perons to describe it in detail. Barr bodies are found within the nuclei of most female cells from many kinds of animals, including humans (Figure 2-11). Nonetheless, Barr bodies are never observed within the nuclei of normal males.

Not long after Barr bodies were first demonstrated in the tissues of normal women, it was reported that women who had Turner's syndrome (XO) did *not* have a Barr body, whereas men who had Klinefelter's syndrome (XXY) *did!* This led to the hypothesis that the Barr body is actually an X chromosome that is

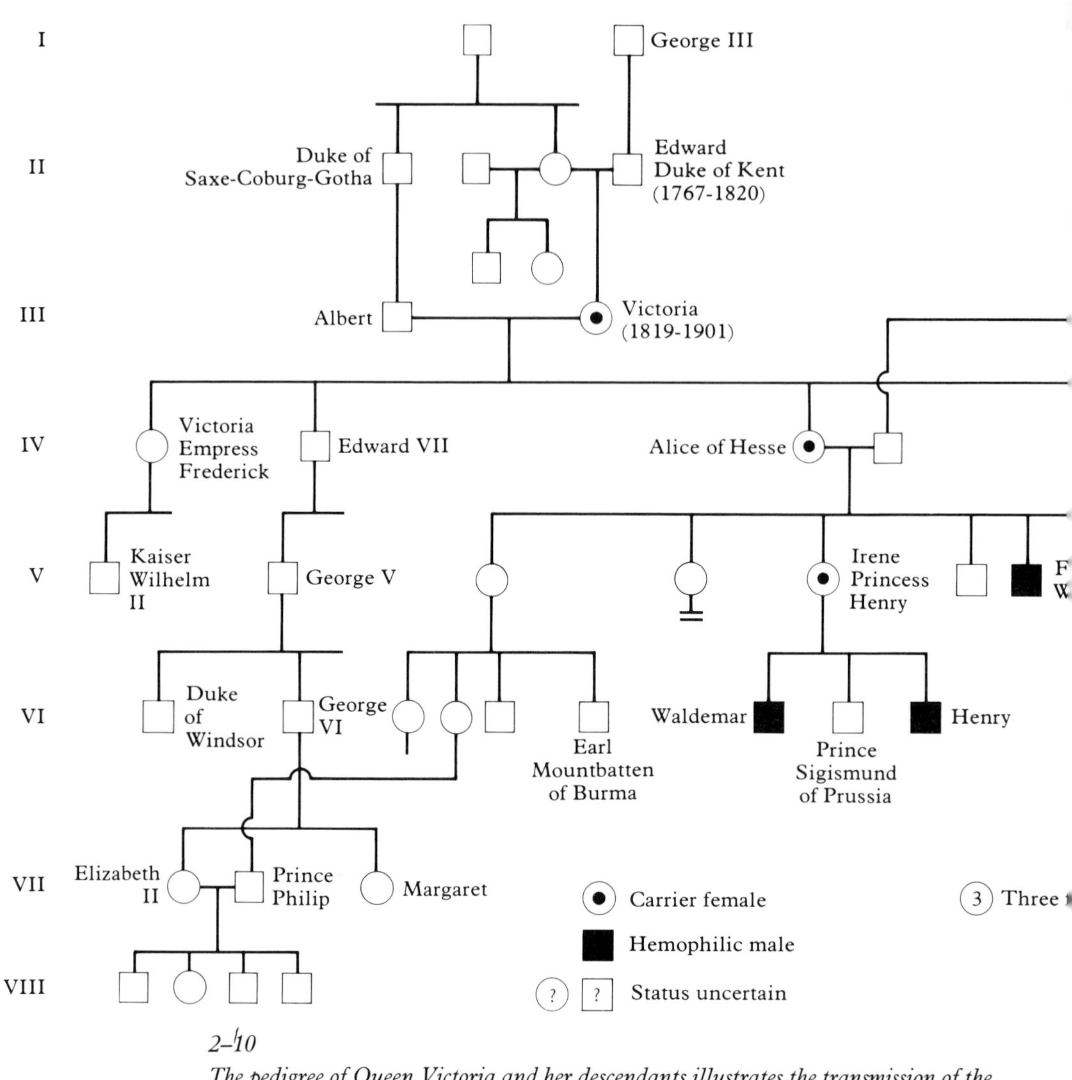

2–10
The pedigree of Queen Victoria and her descendants illustrates the transmission of the X-linked recessive trait hemophilia. (After Dr. Victor McKusick.)

tightly coiled in a dark-staining nuclear blob that is genetically inert. Thus, an obvious explanation for the occurrence of dosage compensation seemed to be this: women and men have the same concentration of gene products determined by X-linked genes because in normal women one X chromosome is genetically inactive.

Further support for the idea that the Barr body is an inactive X chromosome came from the study of the concentration of G6PD in persons who had abnormal sex chromosome constitutions. Thus XO individuals (with no Barr body) and XXY individuals (with one Barr body) both were shown to have roughly normal concentrations of G6PD. Later, it was discovered that individuals whose sex chromosomes are XXX have normal G6PD levels and *two* Barr bodies. Similarly, people who are XXXX have normal G6PD levels and *three* Barr bodies. The number of Barr bodies is thus one less than the total number of X chromosomes, which means that one X chromosome is always available to function normally.

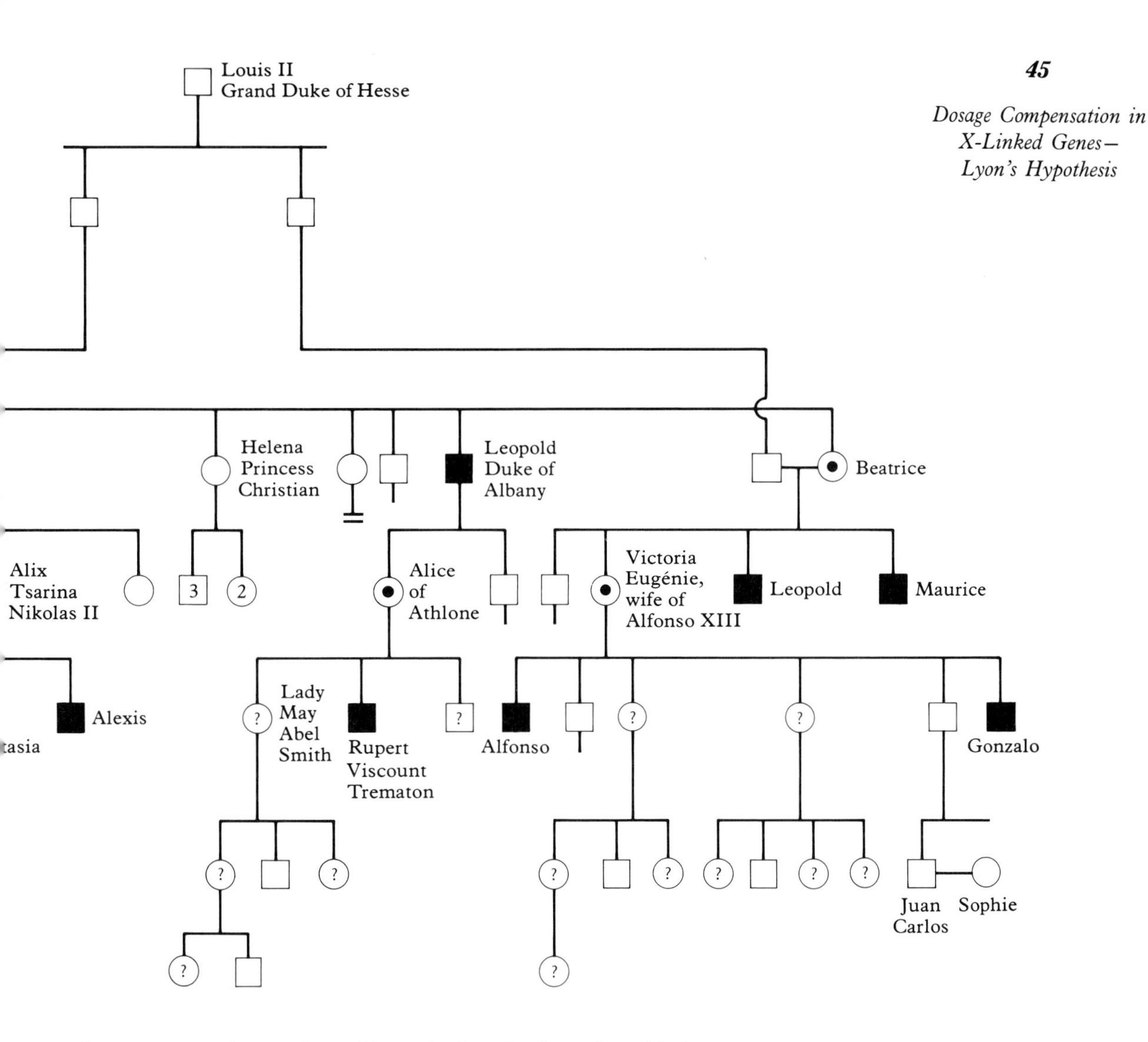

Our present understanding of how the inactivation of one X chromosome relates to dosage compensation is best stated by what has become known as *Lyon's hypothesis,* after the woman who was among the first to propose the idea. (The idea occurred almost simultaneously to several other investigators, and it has been refined over the years as more data have become available.)

The present-day version of Lyon's hypothesis is this. First, very early in embryonic life, when the number of cells in the body of a human female is relatively small, one of the X chromosomes becomes genetically inactive and forms a Barr body. Second, in some cells it is the X chromosome from the mother that is turned off, and in others it is that from the father. In other words, X chromosomes are turned off at random. Third, once the paternal or maternal X chromosome has been turned off in a given cell the same X chromosome is turned off in all of the descendents from that cell during the later development of the embryo. We now turn our attention to some of the evidence that one of the female's X chromosomes really does behave in this rather remarkable way.

If one or the other of the human female's X chromosomes is turned off at random during early fetal development, then one would anticipate that normal women would be mosaics for the X chromosome in that different X chromosomes may be turned on in different cell lines. That this may be true is supported by studies that show that the red blood cells of women who are heterozygous

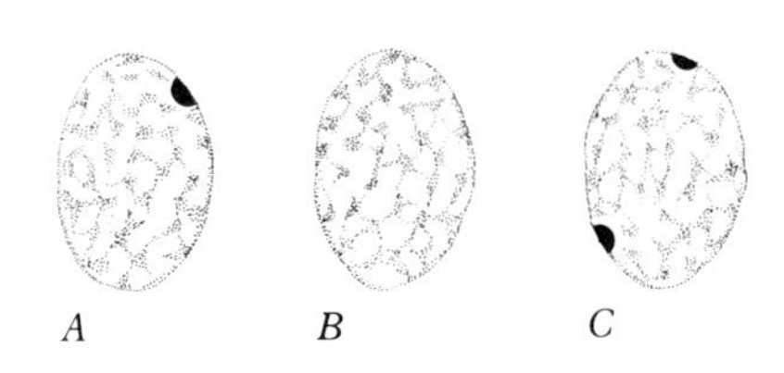

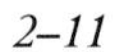

2–11
A, the nondividing nuclei of cells of normal females contain a single Barr body. B, the nuclei of cells of normal males lack a Barr body. C, the nuclei of cells of persons who have sex chromosomes XXX or XXXY have two Barr bodies. Also see Table 2-1.

carriers for G6PD deficiency clearly fall into two types. They have either normal enzyme activity or almost none. Presumably, the two different populations are the descendants of early embryonic cells in which opposite X chromosomes were inactivated.

Further evidence supporting the inactivation of one or the other X chromosome in different female cell lines comes from determinations of the concentration of G6PD in the red cells of women who are affected by G6PD deficiency. As we would expect, most women who are affected by G6PD deficiency are homozygous recessives, which is to say that they have an abnormal allele on each X chromosome. Nonetheless, some G6PD-deficient women turn out to be heterozygous, and although the concentration of G6PD inside their red blood cells is usually intermediate, it may vary from as low as that of homozygous recessive women to as high as that of normal men and women. This makes sense if we assume that most of the red blood cells of heterozygous women with very low concentrations of G6PD are the descendants of embryonic cells in which the X chromosome that had the abnormal allele was the active one. Similarly, the red cells of heterozygous women who have very high concentrations of G6PD are presumably the descendants of embryonic cells that happened to have the X chromosome whose abnormal allele was turned off. (Women have manifested X-linked recessive traits in spite of the fact that they are heterozygous for hemophilia and some other traits. Such women are known as *manifesting heterozygotes;* presumably, most of their normal X chromosomes are inactivated by chance in early embryonic life.)

Although the evidence in favor of the Lyon hypothesis is convincing overall, there are still some unanswered questions. Perhaps the most pressing of them is this. Normal XX females and normal XY males both have only one active X chromosome. Why then is it that those who have Turner's syndrome (XO) and Klinefelter's syndrome (XXY), both of whom also have only one active X chromosome, are not only sterile but distinctly abnormal in several other ways?

It has been suggested that the abnormal phenotypes of XO and XXY individuals may result because inactivity affects most, but not all, of the X chromosome. If the proposed portion of the mostly-turned-off-X chromosome that remains active carries genes that determine the phenotypic differences between XO and XX individuals, then the phenotypic abnormalities of the XO genotype could be accounted for. Although XO individuals have one complete X chromosome, they lack the supposedly active portion of a normal woman's other, mostly inactivated, X chromosome. Similarly, individuals of genotype XXY would be abnormal because they have a normal X, a normal Y, *plus* the active portion of the other, mostly inactive, X chromosome.

Overall, Lyon's hypothesis seems fairly well established, and evidence is accumulating fast enough that it should be possible to accept or reject the hypothesis in the near future.

Summary

For most sexually reproducing animals it is advantageous to have about as many males as females, and these equal numbers are usually maintained by male-female differences in chromosomes.

All normal human beings have 22 pairs of autosomes; in addition, normal women have two X chromosomes, whereas normal men have one X chromosome and one Y chromosome. The sex of an individual is determined at fertilization and depends on whether the egg is fertilized by an X-bearing or a Y-bearing sperm.

The sex ratio of the human population varies with age. More males than females are probably conceived and born, but at every stage of life males are more likely to die than females. Voluntary selection of the sex of offspring may soon be a reality for some populations.

The discovery of the chromosomal basis of Turner's syndrome (XO) and Klinefelter's syndrome (XXY) suggested a male-determining role for the human Y chromosome, and this was borne out by studies of human sexual mosaics. In the absence of the Y chromosome the phenotype is female as long as at least one X chromosome is present. Autosomal genes must also influence sex determination, and both sexes have genes for "maleness" and "femaleness" on their autosomes.

The Y chromosome is nearly devoid of genes not involved in sex determination, but the X chromosome is known to carry at least 100 genes other than those that determine sex. Most X-linked abnormalities are inherited as recessive traits. X-linked traits, such as classical hemophilia, have distinctive patterns of inheritance.

Dosage compensation occurs for X-linked traits because although women have two X chromosomes, they do not have twice the concentration of products of X-linked genes that men, with one X chromosome, have. Lyon's hypothesis suggests that one or the other of the female's X chromosomes is randomly inactivated very early in development. Inactivated X chromosomes can be observed within the nuclei of female cells, and they are called Barr bodies. The number of Barr bodies is always one less than the number of X chromosomes, and in normal women opposite X chromosomes may be active in different cell lines. Normal women are therefore mosaics for traits determined by X-linked genes.

Suggested Readings

1. "Sex Differences in Cells," by Ursula Mittwoch. *Scientific American,* July 1963, Offprint 161. A review of the major chromosomal differences between men and women.
2. *Genetic Mosaics and Other Essays,* by Curt Stern. Harvard University Press, 1968. A short, rather technical work for those especially interested in genetic mosaics.
3. "Sex Preselection in the United States: Some Implications," by Charles F. Westoff and Ronald R. Rindfuss. *Science,* vol. 184, 10 May 1974. What would happen to the secondary sex ratio if sex preselection were freely available in the United States?

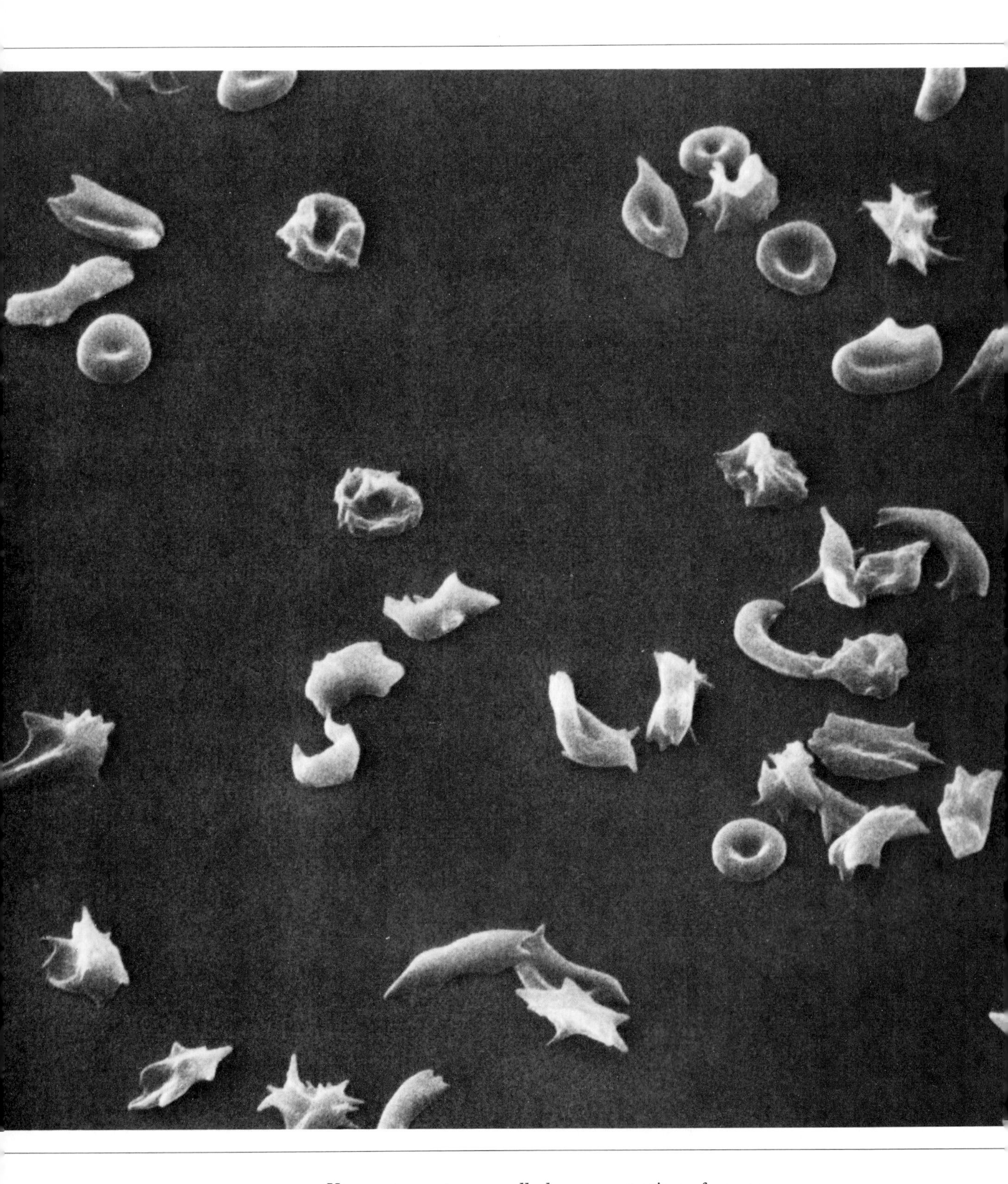

Upon exposure to unusually low concentrations of oxygen, the red blood cells of persons who have sickle-cell trait may become distorted, as shown in this scanning electron micrograph. (Courtesy of Patricia Farnsworth.)

CHAPTER 3

Genes and the Individual

In 1869, only three years after Mendel had published his manuscript, a Swiss biochemist treated some cells with the enzyme pepsin and discovered that although the nuclei of the treated cells shrank in size they were not completely dissolved. Because pepsin is an enzyme that digests protein molecules, it was concluded that nuclei consist, not only of protein, but of other substances, too. Further testing of the undigested nuclear material showed that it was rich in phosphorus and that it could be purified into a white powder. This purified nuclear material eventually became known as *nucleic acid,* and, as was also true of Mendel's manuscript, the scientists of the day took little note of it. The white powder was to spend many years on dusty laboratory shelves before it was realized that it was the hereditary material itself, deoxyribonucleic acid (DNA).

By 1924 specific staining techniques had been devised so that it was possible to see the exact location of DNA within the nucleus. It was discovered that in dividing cells nuclear DNA is always found in chromosomes. It also became known that all of the body cells of a given organism contain the same amount of DNA, whereas the sex cells (eggs and sperm) contain only half as much. This clearly suggested that DNA might be the genetic material, but for some reason the idea was not widely accepted by biologists until the early 1950s.

Before then, most biologists were convinced that the genetic material was not DNA, but protein, which is abundant in both nondividing nuclei and in chromosomes. Proteins were known to be very complicated molecules and what little was then known of the structure of DNA suggested that, although its molecules were very long and threadlike, they were probably simpler than protein molecules. Most biologists felt that protein was more likely to be the genetic material than DNA because complex protein molecules seemed more appropriate bearers of genetic information than the apparently simpler molecules of DNA.

Then in 1944 it was reported that DNA alone, and not protein, could account for a phenomenon known as *transformation* that had been discovered some years earlier. Transformation can be thought of as a lasting change in the genetic program of certain bacteria brought about by DNA from other bacteria. Thus, as shown in Figure 3-1, when purified DNA from dead bacteria of Type 1 is added to living cultures of organisms of Type 2, some *living* organisms that have characteristics of Type 1 may form. This is because some cells of Type 2 take up Type 1 DNA from the culture medium and thus become transformed. Once they are transformed, the organisms breed true to their new type and it is possible to recover more Type 1 DNA from the transformed bacteria than was originally added to the growing culture. Thus not only does Type 1 DNA transform some Type 2 cells into cells that have some of the characteristics of Type 1, but the resulting Type 1 organisms reproduce and manufacture new Type 1 DNA as they do so.

Propagation of Type 1
bacteria (smooth colonies)

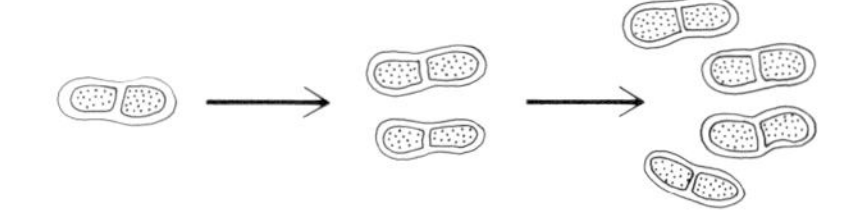

Preparation of transforming principle:

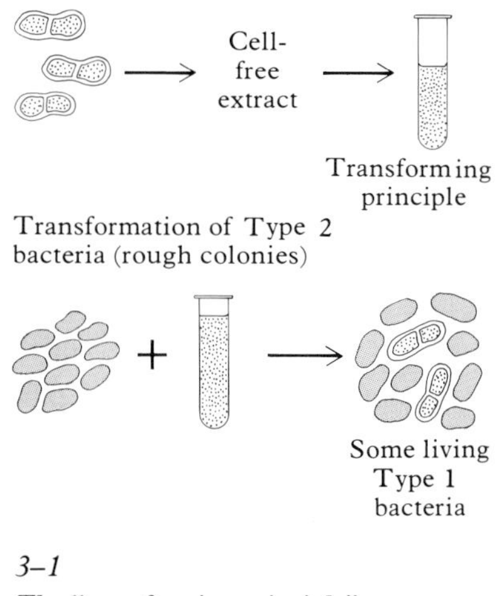

3–1
The "transforming principle" was found to be DNA, not protein. (From Gunther S. Stent, Molecular Genetics. *W. H. Freeman and Company. Copyright © 1971.)*

Further support for the idea that DNA is the genetic material came from the study of certain kinds of *viruses.* In general, viruses are composed of a protein coat surrounding a core of nucleic acid. (The nucleic acid in viruses is usually, but not always, DNA.) In 1952 it was reported that certain kinds of viruses that attack and usually destroy the bacterium *Escherichia coli (E. coli),* a normal inhabitant of the human digestive tract, are able to do so because they inject DNA into the bacterium like tiny parasitic syringes. Such a virus is known as a *bacteriophage,* or *phage* (see Figure 3-2). The phage's protein coat attaches the virus to the bacterium, but, unlike the DNA, the protein coat does not get inside the cell. The DNA injected into the bacterium contains all of the information necessary to manufacture entire new phage particles, including the protein coat.

Thus, by 1952 there was little doubt that DNA was the hereditary material. As we will discuss further in this chapter, it is now known that the protein in the nondividing nucleus and in chromosomes provides a physical support, or scaffolding, for DNA molecules, and plays a part in determining which genes are expressed when. But it is in the enormously long, double-stranded, self-replicating molecule of DNA that we find the final, biochemical basis of heredity.

The unraveling of the biochemical basis of heredity is one of the greatest intellectual achievements of this or any other century. But a detailed description of how our knowledge of biochemical genetics came about, no matter how engaging it may be, is beyond our present purposes. In this chapter, our main concern is with identifying the biochemical basis of some rather well-understood genetic abnormalities of human beings and with how genes bring about their effects in individuals. Our present understanding of human biochemical genetics, indeed of biochemical genetics in general, is in large part an outgrowth of a proposal for the detailed structure of DNA first put forth in 1953. The structure of DNA is so important, and so revealing, that we must discuss it further.

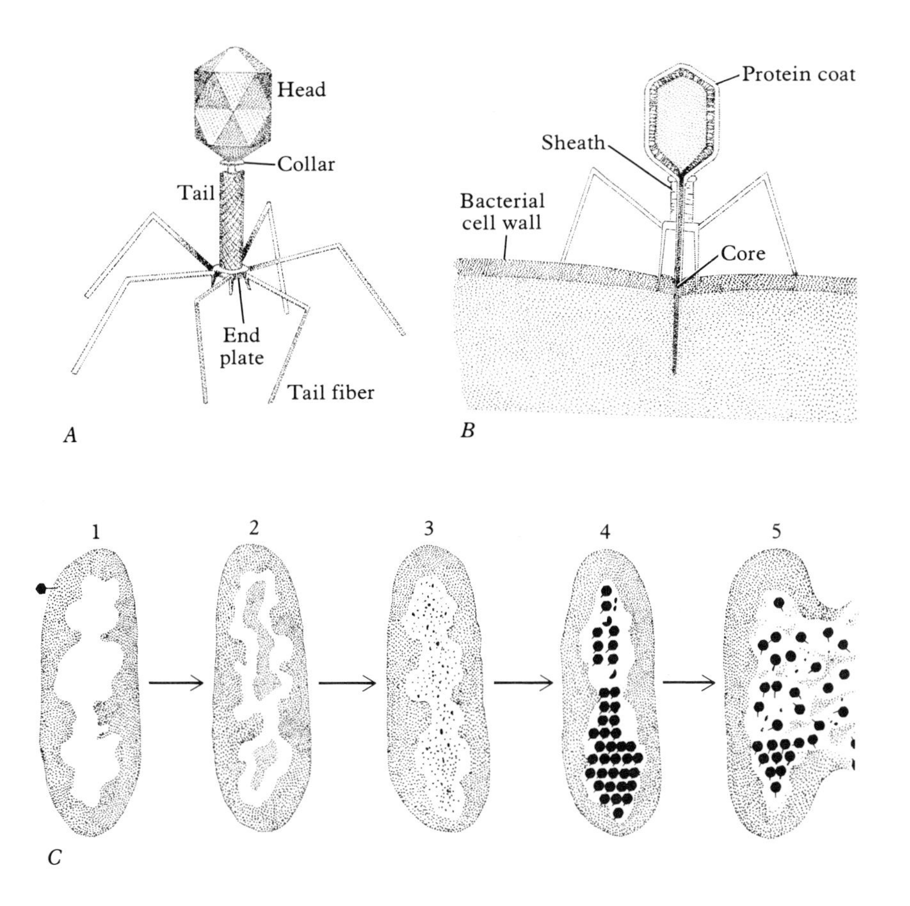

3–2
Phage viruses are parasites of bacterial cells. A, the head of the virus is a protein shell that contains DNA. B, the virus attaches to a bacterial cell and injects viral DNA into it. C, following the injection of viral DNA the bacterial cell is converted into a virus-making factory. The bacterium eventually bursts, releasing the newly formed phage. (From "Building a Bacterial Virus," by William B. Wood and R. S. Edgar. Copyright © 1967 by Scientific American, Inc. All rights reserved.)

The Watson-Crick Model for the Structure of DNA

Like most other very large, naturally occurring molecules, DNA is made up of a few relatively simple chemical building blocks that are joined to one another in sequence by means of chemical bonds. In DNA, these building block compounds are *nucleotides.* Each nucleotide is made up of three parts: a phosphate group, a sugar that contains five carbon atoms and is known as deoxyribose, and a nitrogen-containing base (Figure 3-3).

There are four kinds of nucleotides in DNA. All of them contain the phosphate and the sugar, but they differ in their nitrogen-containing bases. The four bases fall naturally into two categories according to their structure. Two of the bases, *adenine* and *guanine,* have a double-ring structure, and the remaining two bases, *cytosine* and *thymine,* are made up of a single ring, as shown in Figure 3-4.

Within a DNA molecule the nucleotides are bonded in such a way that the sugar of one nucleotide is always attached to the phosphate group of the next nucleotide in line. This arrangement results in a long chain consisting of alternating sugar and phosphate groups, and from this backbone the bases stick out to the side. (Figure 3-5). This was well known before Watson and Crick published their model for the structure of DNA in 1953. Their great accomplishment was not an explanation of the *composition* of DNA, but rather of the detailed three-dimensional architecture of the molecule.

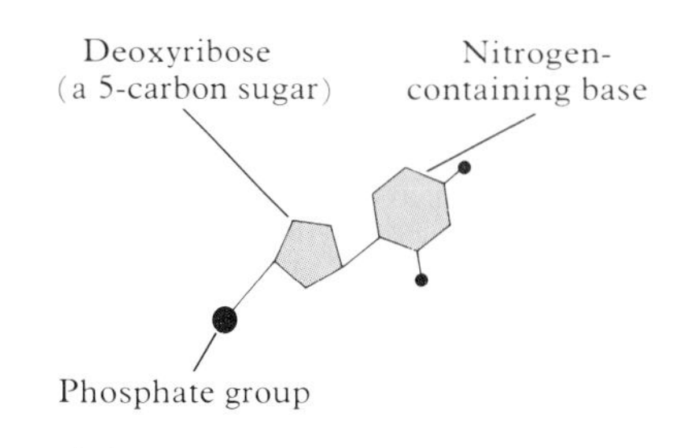

3–3
The structure of a nucleotide, one of the building blocks of DNA.

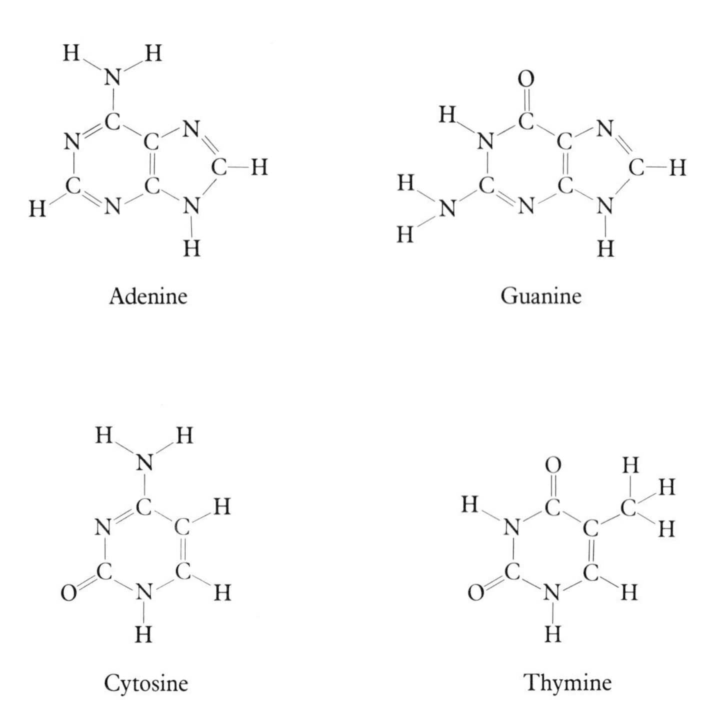

3–4
The structural formulas of the four nitrogen-containing bases that are found in DNA.

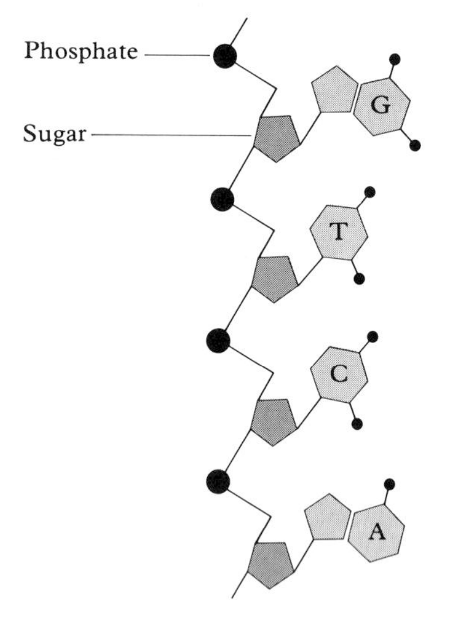

3–5
The backbone of the DNA molecule is made up of alternating sugar and phosphate groups. The nitrogen-containing bases stick out from the backbone. Adenine (A) and guanine (G) are the double-ring bases; thymine (T) and cytosine (C), the single-ring bases.

The model was based on information from several sources. It had been reported a few years earlier that all DNA molecules, no matter what their source, have something in common in the number of bases they possess. Although the amounts of the four bases may vary widely from species to species, in all DNA molecules the number of adenine bases is exactly equal to the number of thymine bases, and the number of guanine bases is exactly equal to the number of cytosine bases.

Watson and Crick used this information and results from X-ray diffraction studies and from other experiments aimed at determining the exact distances between atoms in the DNA molecule and devised a model of DNA structure that is simple and elegant. The overall model can be summed up in the following way.

First, the DNA molecule is a double helix. To visualize a helix, think of the backbone of alternating sugar and phosphate groups as wrapped around a long thin cylinder. In the double helix of DNA, two backbones (two strands) are present, and they are held together because the bases stick out into the interior of the molecule in a way that enables them to form weak chemical bonds with one another. (See Figure 3-6.)

Second, the amount of adenine is equal to that of thymine because an adenine on one strand is always bonded to a thymine on the opposite strand. Similarly, the amount of guanine is equal to that of cytosine because in DNA these two bases are always bonded to one another across the double helix.

One of the most convincing points of this model was that it immediately suggested how DNA might replicate itself. Because adenine always pairs with thymine and guanine always pairs with cytosine, if the two strands of a DNA molecule are separated by breaking the bonds between the bases, then each chain provides all the information necessary to synthesize a new partner. It was soon discovered that DNA is indeed capable of self-replication, as any molecule reputed to be genetic material must be. As shown in Figure 3-7, DNA

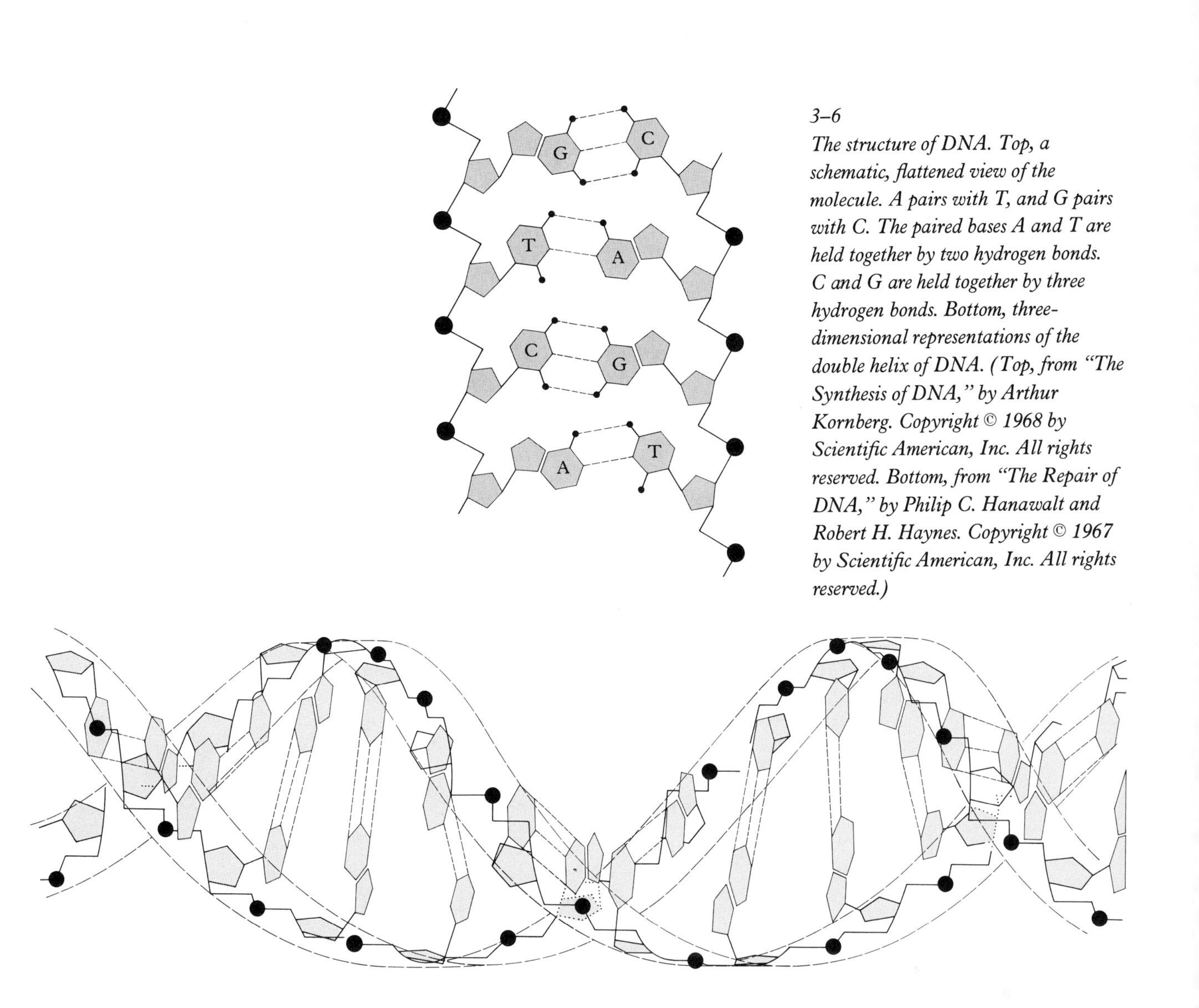

3–6
The structure of DNA. Top, a schematic, flattened view of the molecule. A pairs with T, and G pairs with C. The paired bases A and T are held together by two hydrogen bonds. C and G are held together by three hydrogen bonds. Bottom, three-dimensional representations of the double helix of DNA. (Top, from "The Synthesis of DNA," by Arthur Kornberg. Copyright © 1968 by Scientific American, Inc. All rights reserved. Bottom, from "The Repair of DNA," by Philip C. Hanawalt and Robert H. Haynes. Copyright © 1967 by Scientific American, Inc. All rights reserved.)

replication is accomplished by the separation of the two strands of the double helix. Each strand then serves as a template for the manufacture of a new complementary strand.

But there remained an obvious question about the Watson and Crick model. DNA had been proved to be the hereditary material, but nobody yet knew how the double helix brings about its biochemical effects. Nonetheless, within a few years laboratories from all over the world had contributed to an overall biochemical scheme of gene action whose details are still being worked out today. In sum, at the biochemical level genes usually participate in *protein synthesis,* and within the double helix is encoded all of the information necessary for a cell to synthesize the proteins it needs to survive and reproduce.

As shown in Figure 3-8, the relationship between DNA and protein synthesis can be summed up in this way. Proteins are made up of building blocks called *amino acids,* of which there are at least 20 kinds. The number of amino acids in a protein molecule may vary from about 50 to 50,000 or more, and in general the properties of a particular protein depend above all on the sequence of amino acids that makes up the molecule (Figure 3-9).

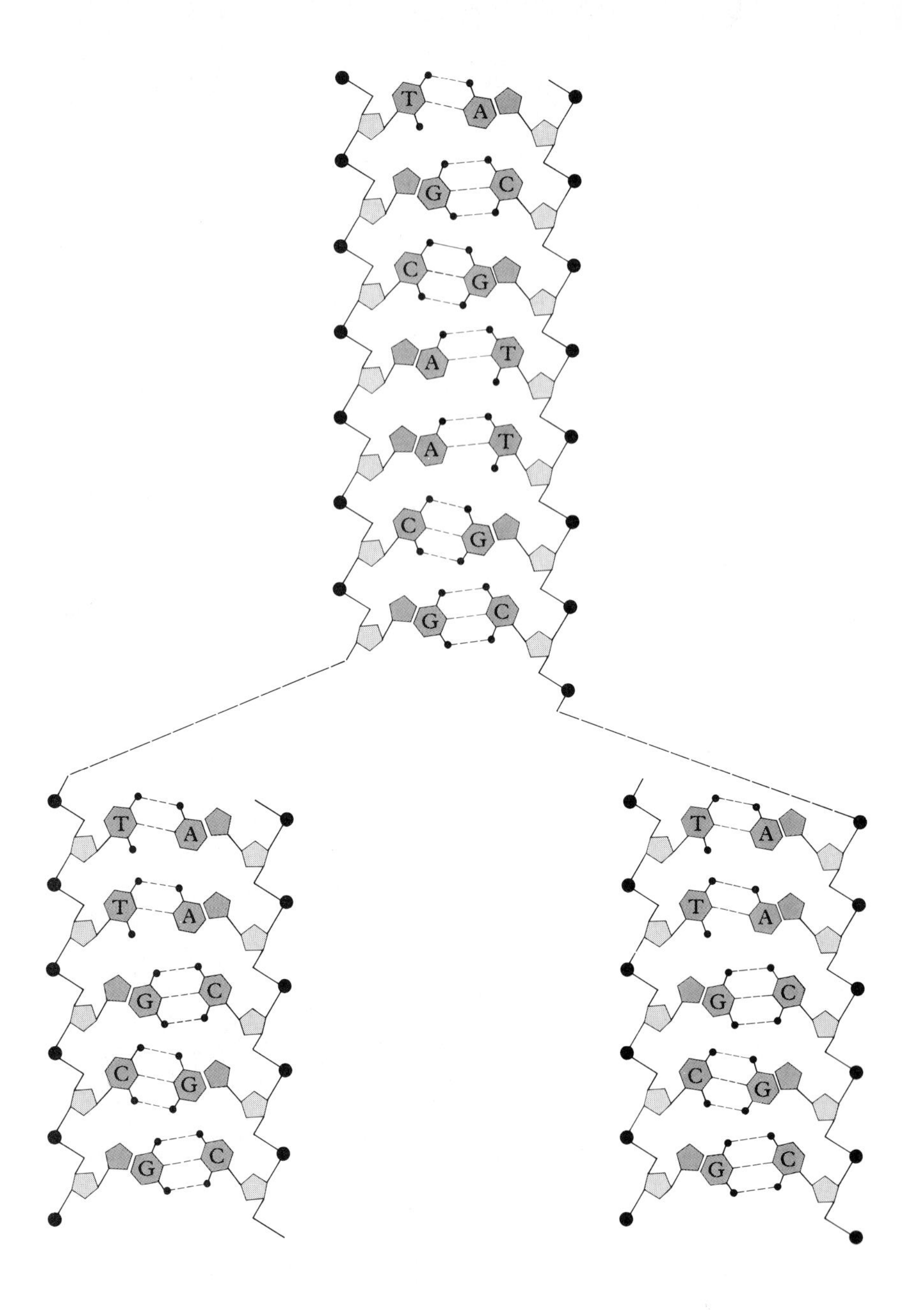

3–7
During DNA replication the two strands of the double helix separate and each of them serves as a template for the manufacture of a new complementary strand. (From "The Synthesis of DNA," by Arthur Kornberg. Copyright © 1968 by Scientific American, Inc. All rights reserved.)

What determines the sequence of amino acids in a given protein is the sequence of the bases in a particular segment of the molecule of DNA. As shown in Figure 3-8, *three* bases in the DNA molecule usually code for *one* amino acid.

Protein synthesis also requires a second kind of nucleic acid known as ribonucleic acid (RNA). In RNA the five-carbon sugar is ribose, and RNA does not contain thymine but rather a different single-ring base that pairs with adenine, *uracil.* Also, unlike DNA, most RNA is single stranded. That is, the backbone of RNA is usually a single chain of alternating sugar and phosphate groups.

Protein synthesis begins when a segment of DNA within the nucleus becomes unwound and one of the strands provides the code for the synthesis of *messenger RNA* (mRNA). mRNA then diffuses out of the nucleus into the cytoplasm of the cell where it becomes attached to structures called ribosomes.

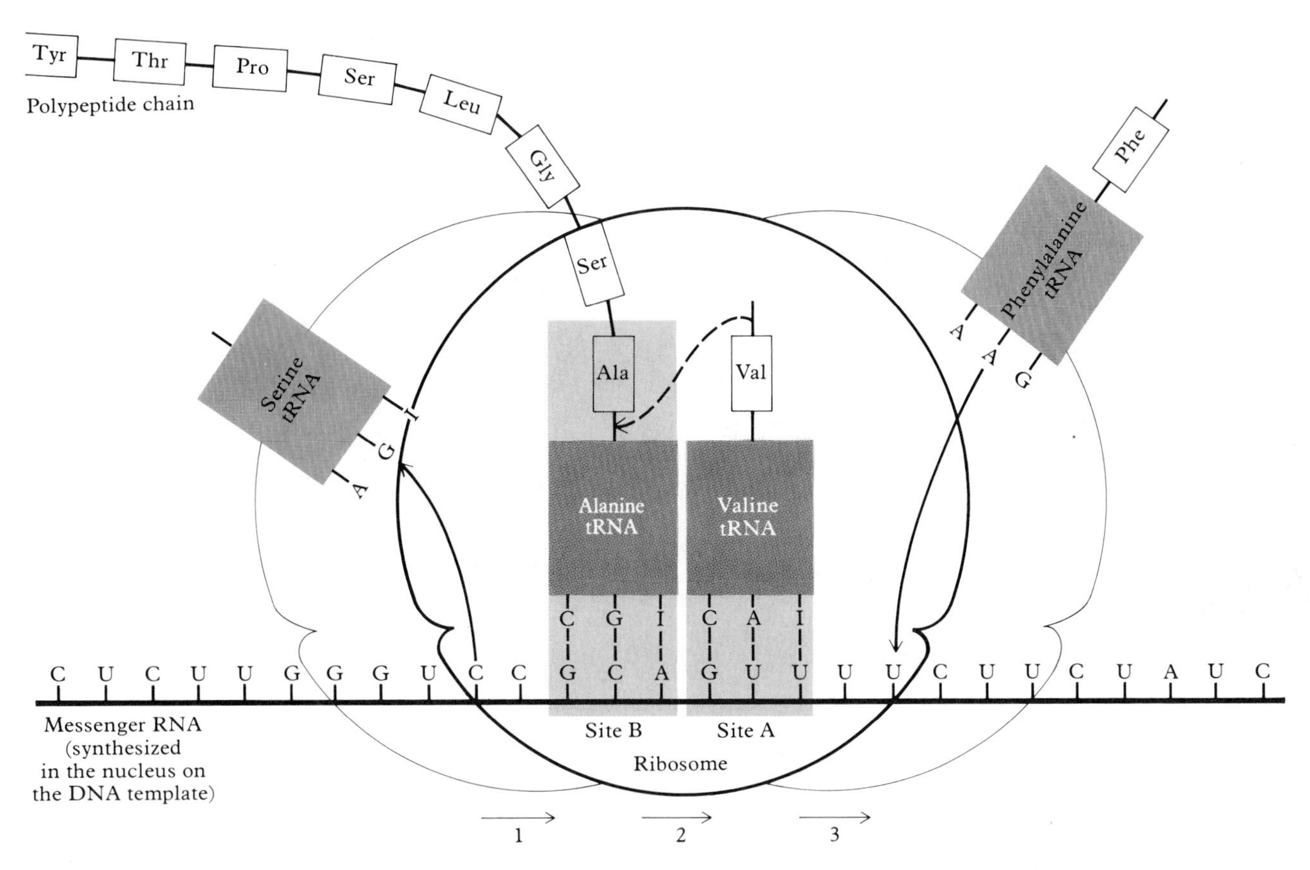

3–8
Protein synthesis occurs on the surfaces of tiny intracellular particles called ribosomes. A ribosome moves along a molecule of mRNA that is complementary to a portion of one of the strands of DNA in the nucleus. On the surface of the ribosome the sequence of bases in the mRNA is translated into a chain of amino acids. Each type of amino acid is carried to the ribosome by a specific tRNA molecule, one portion of which contains three exposed bases. The three exposed bases of tRNA form temporary bonds with a complementary three-base sequence in the mRNA molecule. When this happens, one amino acid is added to the growing chain and the tRNA is released to once again pick up another molecule of its specific amino acid. (Some tRNA molecules contain bases other than those we have already discussed, but each still pairs with a specific base in the mRNA molecule.) (From "The Genetic Code: III," by F. H. C. Crick. Copyright © 1966 by Scientific American, Inc. All rights reserved.)

Ribosomes are the site of the stringing together of specially activated amino acids to form the long chains of protein molecules. On the surface of ribosomes, the base sequence of mRNA, which is complementary to the copied strand of DNA in the nucleus, becomes *translated* into a chain of amino acids. Amino acids are first activated by being attached to another kind of RNA, *transfer RNA* (tRNA). Each amino acid has its own tRNA, and each tRNA contains three exposed bases that can pair with the complementary three bases on the molecule of mRNA. When they do so, tRNA molecules release their particular activated amino acid to the growing amino acid chain. (See Figure 3-8.)

Thus, at the biochemical level, most genes turn out to be specific regions along enormously long DNA molecules that code for the manufacture of a

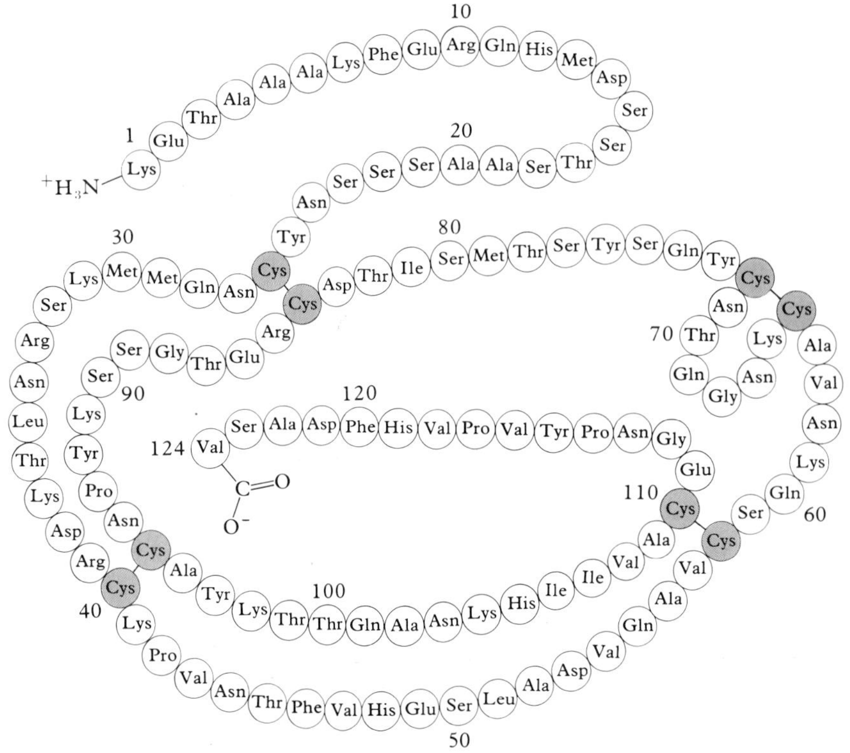

Amino acid	Three-letter abbreviation
Alanine	Ala
Arginine	Arg
Asparagine	Asn
Aspartic acid	Asp
Asparagine or aspartic acid	Asx
Cysteine	Cys
Glutamine	Gln
Glutamic Acid	Glu
Glutamine or glutamic acid	Glx
Glycine	Gly
Histidine	His
Isoleucine	Ile
Leucine	Leu
Lysine	Lys
Methionine	Met
Phenylalanine	Phe
Proline	Pro
Serine	Ser
Threonine	Thr
Tryptophan	Trp
Tyrosine	Tyr
Valine	Val

3–9
The properties of a particular protein depend above all on the sequence of its component amino acids. This is the molecule of bovine ribonuclease, an enzyme that digests RNA. The chemical bonds that form between adjacent cysteine help to determine the shape of the enzyme molecule.

particular protein by the mechanism just described. How does all of this relate to people? To begin with, we can use this information concerning the biochemical bases of protein syntheses to form a rough estimate of what the total number of human genes may be.

DNA and the Structure of Chromosomes

Because most genes can be thought of as sections of DNA molecules that contain coded messages for the manufacture of specific proteins, we can use the total amount of DNA inside the cells of various organisms to estimate how many genes the organisms have. As we have seen, three successive bases in the DNA molecule code for one amino acid. If we assume that an average protein molecule consists of about 200 amino acids (a reasonable estimate), this means that about 600 successive bases in DNA are required to code for an average protein molecule. So if we know the total number of base pairs in the DNA of a particular species, we can estimate the number of genes that code for proteins by dividing the total number of base pairs by 600.

When these calculations are carried out the results are often surprising. For example, ordinary toads are estimated to have about 7 million genes, whereas frogs have 14 million. The same method gives an estimate for human beings of about 5 million genes, and the estimate for a certain species of salamander is a whopping 125 million genes! It is not at all obvious why a frog should have twice as many genes as a toad, or why a salamander should have 25 times as many genes as a human being. On the contrary, most biologists agree that the number of different kinds of protein molecules that have structural roles or that influence metabolism probably does not vary much, perhaps tenfold, from the lowliest multicellular creatures to the most complicated. How then can we account for the widespread variation in the amount of DNA observed from one multicellular creature to another?

At least part of the answer is found in the recent discovery that in most multicellular organisms some segments of the DNA molecule are present in multiple copies. In fact, a significant proportion of the DNA in most multicellular organisms consists of segments that have similar or even identical base sequences, some of them repeated thousands, or as many as a million, times. The genetic role of these widespread duplicated segments of DNA is far from clear, but they can account, at least in part, for the rather extreme variation in the amount of DNA from one species of multicellular organism to another.

In the end, we do not know very much about the exact relation between the amount of DNA and the number of genes in any multicellular creature. According to some experts, a good guess of the "actual" number of human genes, taking duplications and other factors into account, may be somewhere between 25,000 and 100,000 genes. In large part, the reason for our relative ignorance of the workings of DNA in multicellular organisms is that most of what has been learned of biochemical genetics so far has come from the study of viruses and bacteria.

As you may recall, viruses are composed of a protein coat surrounding a nucleic acid core. The core usually consists of a short DNA molecule, though some exceptional viruses contain double-stranded RNA instead. Bacteria are far more complicated than viruses and they contain more DNA. Nonetheless, bacteria are composed of *prokaryotic* cells that are much simpler than the *eukaryotic* cells of higher organisms that we have been discussing. The bacterial cell lacks a distinct nucleus; its chromosome (which is never visible in the light microscope) consists of a single DNA molecule and is in the form of a circle (see Figure 3-10). In general, the bacterial chromosome is present in a single copy.

It is a very large step indeed from understanding the organization and workings of the bacterial chromosome to understanding how the genetic material is organized and how it functions inside the nuclei and chromosomes of eukaryotic cells. However, it is known that the mechanism of protein synthesis is more or less the same in both prokaryotic and eukaryotic cells. It has also been shown that genes known as *regulator genes,* which were first reported from bacterial cells, are also present in the DNA of eukaryotes. These genes do not code for a particular structural protein, but rather play a role in *regulating* protein synthesis. We will discuss them in more detail later in this chapter. Overall, most of what has been learned from bacterial genetics can be applied to the genetics of people or any other eukaryotic organisms, but it can tell us little about how the genetic material in eukaryotes is organized. Nonetheless, there have been some recent advances in our understanding of how DNA is related to the complicated chromosomes of eukaryotic cells. These discoveries are worth discussing further.

By weight, eukaryotic chromosomes contain about equal amounts of protein and DNA. Recent advances in our understanding of eukaryotic chromosomes have for the most part centered on the relationship between DNA and chromosomal proteins. The nuclear proteins known as *histones* have at least two important functions. First, they provide a scaffold for the DNA molecules that is most visible when the nuclear contents condense into chromosomes

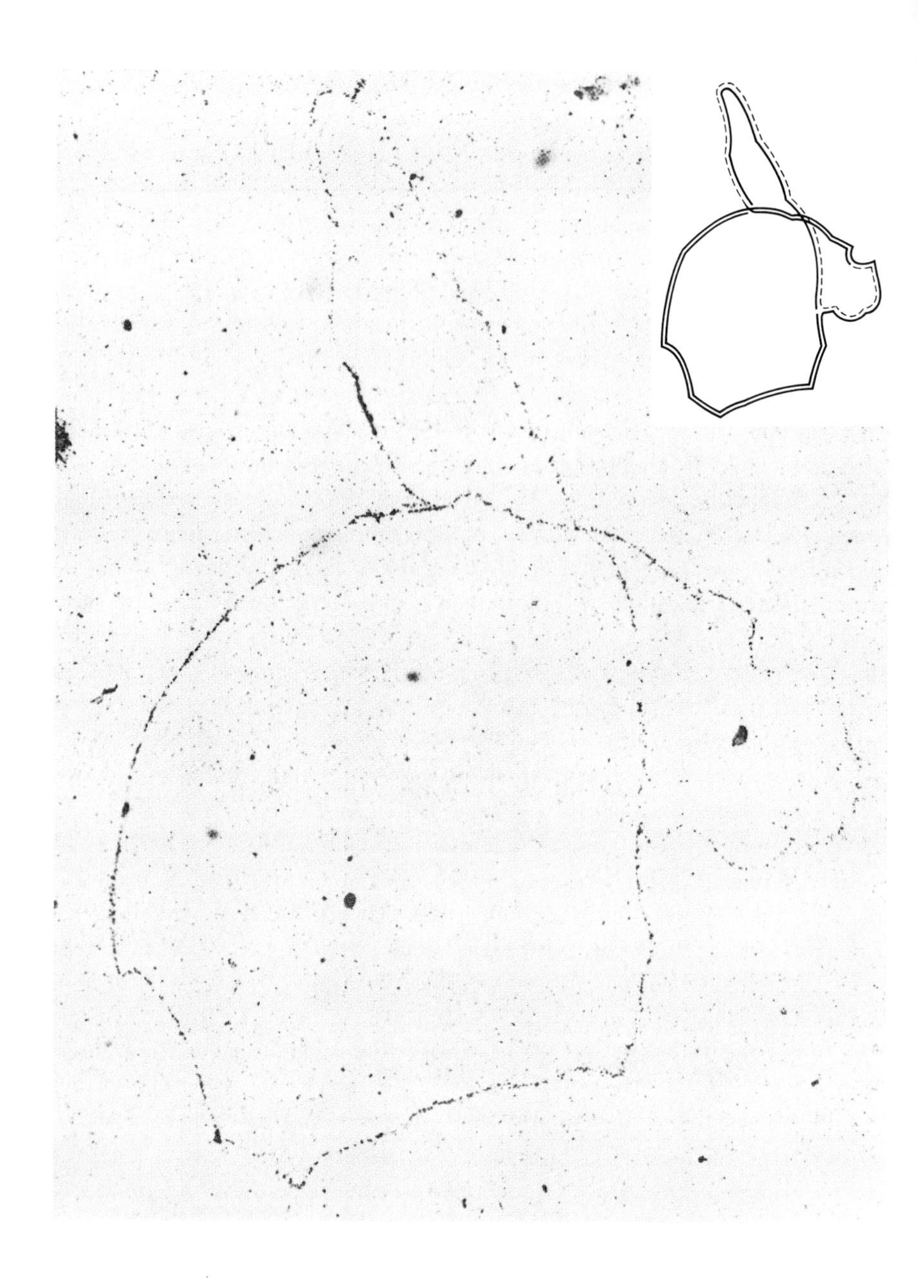

3–10
A bacterial DNA molecule in the act of replicating itself. In the small inset the unduplicated portion of the molecule is indicated by a dotted line. (Courtesy John Cairns.)

during the process of cell division. The central scaffolding may rather resemble beads on a string, with the DNA wrapped around the outside of the beaded structure. Second, histones, along with other proteins in the nucleus, play a major role in regulating the activity of different segments of the DNA molecule. It has been estimated that 10 percent or less of the total genetic material of any eukaryotic cell is biochemically active at any one time, and the nuclear proteins determine which genes are expressed when, though how they do this remains far from clear.

We know very little about the fine structure of the chromosomes of the human species. For example, we do not even know how the total amount of DNA inside human cells is divided among the individual chromosomes. It is possible, though unproved, that the DNA in human chromosomes exists as a very long unbroken thread. For the longest human chromosome, this thread, when completely uncoiled, would be about six centimeters (2.4 inches) long. Obviously, even when not completely unwound, this enormously long molecule would have to be highly folded or coiled within the microscopic nucleus. When eukaryotic cells are not undergoing division, their chromosomes are stretched and loosely entangled with one another inside the nucleus, but the chromosomes are still intact. It is in this relaxed and stretched condition that DNA engages in protein synthesis. On the other hand, the molecular events of crossing over apparently do not occur until the early stages of meiosis, at which time the chromosomes are just beginning to condense into easily distributable packets of genetic information.

With this discussion of biochemical genetics and the structure of eukaryotic chromosomes as background, we now turn our attention once again to the genetics of people. In particular, we can now discuss a famous and frustrating human disease known as *sickle-cell anemia,* which is the result of an abnormality in the genetic code.

Structural Genes–Sickle-Cell Anemia

The clinical syndrome known as sickle-cell disease was first described in 1910. At that time, few facts were known about it. First, those who were affected were blacks of both sexes, and many of them died of it in early childhood. Second, the disease was characterized by the occurrence of sometimes fatal crises that usually lasted for a few days at a time. During a crisis an affected person would develop fever and experience intense and incapacitating pain in the bones, large joints, abdomen, and elsewhere. Third, some of the red blood cells of affected persons were shaped like crescents, or sickles, and all those who were affected had severe anemia. That is, the total number of red blood cells in their circulation was much less than normal (Figure 3-11).

A few years later it was discovered that the sickling of the red cells is related to the state of oxygenation of the iron-containing pigment, *hemoglobin.* Hemoglobin molecules inside red blood cells pick up oxygen in the lungs and release it to the tissues. Red blood cells from persons who have sickle cell disease look normal when their hemoglobin molecules are saturated with oxygen. But when the saturation of hemoglobin by oxygen falls to lower than normal, sickling occurs; and the process can be observed through the microscope. As the con-

3–11
As compared with normal hemoglobin, the abnormal hemoglobin molecules of people who have sickle-cell disease and sickle-cell trait (to be discussed later) show distinctive patterns of movement in an electric field. The areas under the curves correspond to the position and amounts of hemoglobin after an electric current has passed through the liquid in which the hemoglobins are dissolved. The black arrow indicates the location of the specimen before the electric field was applied.

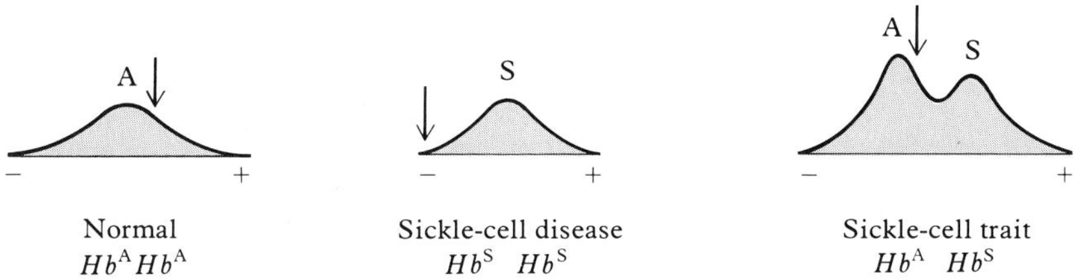

centration of oxygen falls, more and more of the "normal" cells turn into sickle cells right before one's eyes.

What is it about the red blood cells of persons who have sickle-cell disease that results in their unusual change in shape in the presence of low concentrations of oxygen? The first clue to this question came in 1949 when it was discovered that the hemoglobin of those who have sickle-cell disease differs from normal hemoglobin. As shown in Figure 3-11, sickle-cell hemoglobin molecules were found to have a different pattern of movement than normal hemoglobin molecules in the presence of an electric field. This means that the two molecules cannot be identical. How are they different?

A refinement of the use of an electric field has provided the answer. Hemoglobin can be partially digested by enzymes that split the protein part of the molecule into relatively short amino acid chains of varying length. If the resulting mixture of short amino acid chains is allowed to move under the influence of an electric field, a pattern, or "fingerprint," of the hemoglobin molecule is formed. Each spot of the fingerprint represents a different short chain of amino acids. When this is carried out for normal and sickle-cell hemoglobin, the patterns shown in Figure 3-12 result. Notice that protein digestion results in the formation of 26 short amino acid chains for both kinds of hemoglobin. In sickle-cell hemoglobin, shown on the right in Figure 3-12, 25 of the 26 chains are identical to those in normal hemoglobin. Sickle-cell hemoglobin and normal hemoglobin differ in only one short amino acid chain, that labeled "4" in Figure 3-12.

The exact nature of the difference between sickle-cell and normal hemoglobin is now known. As shown in Figure 3-13, the hemoglobin molecule contains two protein building blocks, the *alpha* chain and the *beta* chain. Each alpha chain consists of 141 amino acids and each beta chain contains 146 amino acids. A normal hemoglobin molecule consists of two alpha chains, two beta chains and four nonprotein, iron-containing portions known as *heme.* As we have just discussed, when the hemoglobin molecule is partially digested by

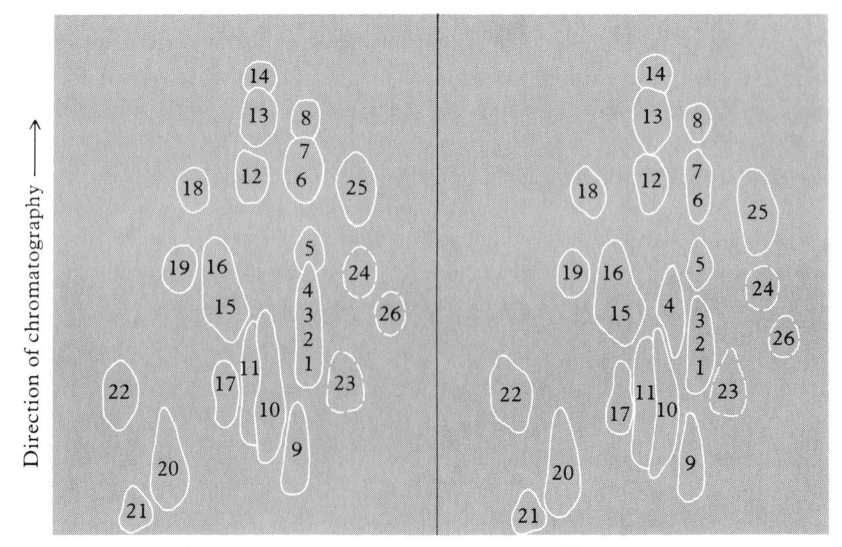

3–12
A "fingerprint" of normal and sickle-cell hemoglobins. Each spot on the fingerprint represents a short chain of amino acids.

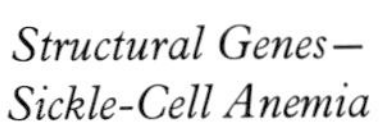

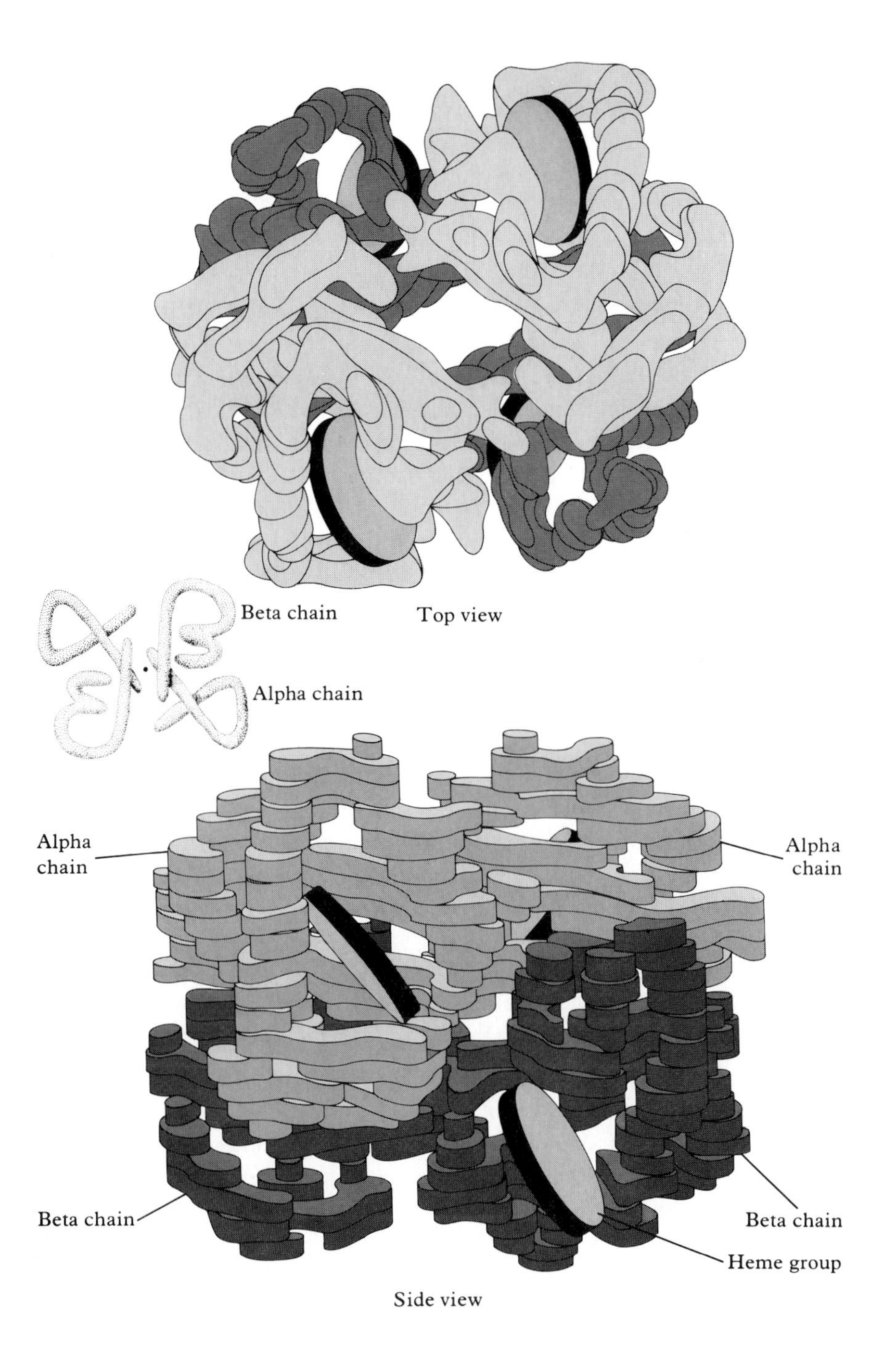

3–13
The hemoglobin molecule consists of two alpha chains, two beta chains, and four heme groups. (From "The Hemoglobin Molecule," by M. F. Perutz. Copyright © 1964 by Scientific American, Inc. All rights reserved.)

protein-splitting enzymes, 26 shorter amino acid chains are left and all but one of these shorter chains (chain 4) are identical in normal and in sickle-cell hemoglobin. That chain on the fingerprint is a short chain consisting of eight amino acids that makes up one of the ends of the beta chain of hemoglobin. Of these eight amino acids, seven are identical in both kinds of hemoglobin. *Only one amino acid* out of a total of 287 is different in sickle-cell hemoglobin. Nonetheless, for persons who have sickle-cell disease, this seemingly trivial molecular difference can mean the difference between life and death.

As you may recall, three base pairs in DNA usually code for one amino acid in a protein molecule. In normal hemoglobin, the three-base sequence CTT (cytosine, thymine, thymine) codes for the amino acid *glutamic acid.* But in sickle-cell hemoglobin, the corresponding three-base sequence is CAT (cytosine, adenine, thymine), which codes for the amino acid *valine.* As shown in Figure 3-14, the amino acid substitution occurs sixth in line from one of the ends of the beta chain.

How does the presence of only one different amino acid out of 287 result in sickle-cell disease? It has recently been shown that the single amino acid difference has no effect on the stability of individual hemoglobin molecules; nor does it alter the molecule's oxygen-carrying ability. Rather, the single amino acid difference results in a unique reaction *between* individual molecules of sickle-cell hemoglobin. Deoxygenated molecules of sickle-cell hemoglobin spontaneously come together to form spiral, rigid, fiberlike structures that distort the red cell and result in sickling. (Most of the red cells quickly resume their normal shape when the hemoglobin is reoxygenated.) The rigid sickle cells tend to become trapped and broken when they circulate through capillaries, and the membrane of the red blood cell may be damaged in the sickling process. Both of these factors contribute to the breakdown of sickle cells that then leads to the anemia that characterizes the disease.

At the present time, many of those who are affected by sickle-cell disease survive beyond the age of fifty. But the improvement in outlook is not due to advances in our understanding of the exact biochemical nature of the disease. For all of our knowledge, we can at present do nothing to alter the composition of the DNA of those who are affected or to alter the translation of DNA into molecules of sickle-cell hemoglobin. Rather, the increased length of survival of affected individuals is largely the result of better nutrition and the prevention of infections that may reduce the amount of oxygen available to the tissues and thus induce widespread sickling.

Shortly after the condition was first described in 1910 it was realized that sickle-cell disease is transmitted as an autosomal recessive trait. This means that affected individuals are homozygous for the gene that results in the production of the abnormal beta chain in the hemoglobin molecule. The genotype of persons who have sickle-cell disease can be represented as $Hb^{S}Hb^{S}$. On the other hand, unaffected individuals are homozygous for normal hemoglobin, or Hemoglobin A, and their genotype is $Hb^{A}Hb^{A}$. People who are *heterozygous* for sickle-cell hemoglobin are thus usually of genotype $Hb^{A}Hb^{S}$. Individuals of genotype $Hb^{A}Hb^{S}$ are said to have *sickle-cell trait* (not sickle-cell disease). Those who have sickle-cell trait do not have anemia and are perfectly normal under most circumstances. Nonetheless, some of the red cells of persons who have sickle-cell trait can be made to sickle in the laboratory by subjecting the cells to lower-than-normal concentrations of oxygen. As shown in Figure 3-11, persons who have sickle-cell trait can also be identified if samples of their hemoglobin molecules are allowed to move in the presence of an electric field. As we might expect, those persons who have sickle-cell trait usually have about equal amounts of normal and sickle-cell hemoglobins throughout their red cells.

Sickle-cell trait occurs in about 8 percent of blacks in the United States, most of whom are of African ancestry. Sickle-cell disease, on the other hand, is becoming less common in the United States, but it still occurs. At least some of the decrease in the number of new cases reported per year is probably related to increased public awareness of the disease and to the detection of normal-appearing individuals who have sickle-cell trait (see Chapter 5). But black

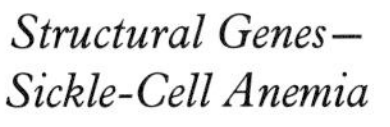

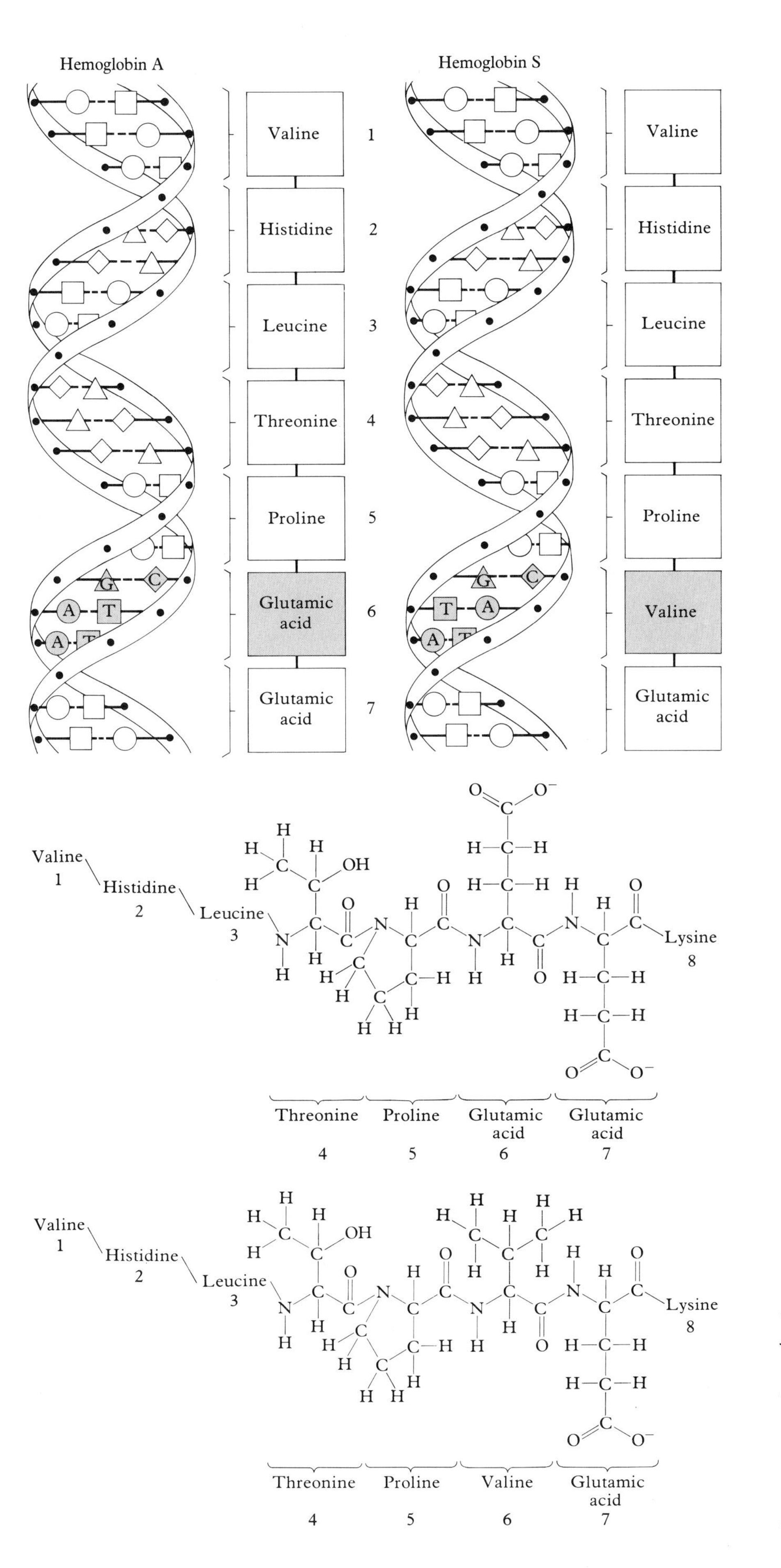

3–14
Only one amino acid out of a total of 287 differs in sickle-cell hemoglobin from normal hemoglobin. As shown here, the amino acid valine replaces glutamic acid. The substitution occurs sixth in line from one of the ends of the beta chain. (From "The Genetics of Human Population," by L. L. Cavalli-Sforza. Copyright © 1974 by Scientific American, Inc. All rights reserved.)

people are not the only human beings afflicted by sickle-cell disease. The abnormal gene also occurs relatively commonly in people inhabiting the Mediterranean area, Arabia, and India. Clearly, people who have sickle-cell disease are at a disadvantage compared to those who have normal hemoglobin. Why, then, is the gene that in the homozygous condition results in sickle-cell disease so widespread in different areas of the world, particularly Africa?

Figure 3-15 is a map showing the distribution of the sickle-cell gene in various parts of the Old World. The highest frequencies are in a rather broad belt across the African continent in which a severe form of *malaria* is a common cause of death. It has been found that the red blood cells of those who have sickle-cell trait ($Hb^{A}Hb^{S}$), which contain a mixture of normal and sickle-cell hemoglobin, are *more resistant to malaria* than normal red blood cells. This advantage probably accounts in large part for the persistent widespread distribution of a gene that in the homozygous state results in reduced reproductive fitness and may lead to early death.

In the past two decades the number of different types of abnormal human hemoglobins reported has increased enormously, and is still doing so. In fact, the deciphering of the biochemical basis of sickle-cell disease and the research that has followed it have brought a mushrooming of our knowledge of molecular disease, which we will discuss at greater length later in this chapter. But the investigation of abnormal hemoglobin molecules has done more than provide us with an understanding of how molecular changes result in disease. It has also provided a clue about how at least some human genes may be regulated. In particular, we have learned that the hemoglobin of normal human beings is different at different stages of development. That is, the kind of hemoglobin present in a normal fetus changes several times during its intrauterine existence, and it is only after birth that Hemoglobin A ($Hb^{A}Hb^{A}$) comes to predominate. Let us discuss this important discovery further.

Fetal Hemoglobins and Regulator Genes

As you know, the protein portion of a normal hemoglobin molecule usually consists of two alpha chains and two beta chains. Each of these amino acid chains is coded for by a different region of nuclear DNA, that is, by a different gene. The gene that codes for the alpha chain becomes active very early in fetal development and remains so throughout adult life. But the gene that codes for the beta chain behaves differently. Although two beta chains are present in almost all of the hemoglobin molecules of normal adults, they are not present in the hemoglobin molecules of a developing fetus or of a newborn infant. The gene that codes for the beta chain becomes active during the second month of embryonic life, and some hemoglobin molecules made up of two alpha chains and two beta chains are present from then on. (As discussed in following paragraphs, in normal adults such molecules of Hemoglobin A compose about 97 percent of the total hemoglobin.) Like the molecules of normal adults, the hemoglobin molecules of developing fetuses and of infants up to six months old contain two alpha chains, but most do not contain two beta chains. Instead of beta chains, most of the hemoglobin molecules of fetuses and newborn infants contain a protein known as the *gamma chain.* This type of hemoglobin is known as *fetal hemoglobin.*

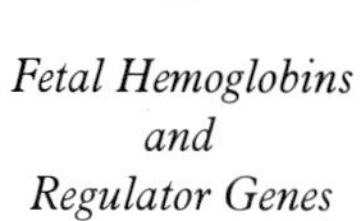

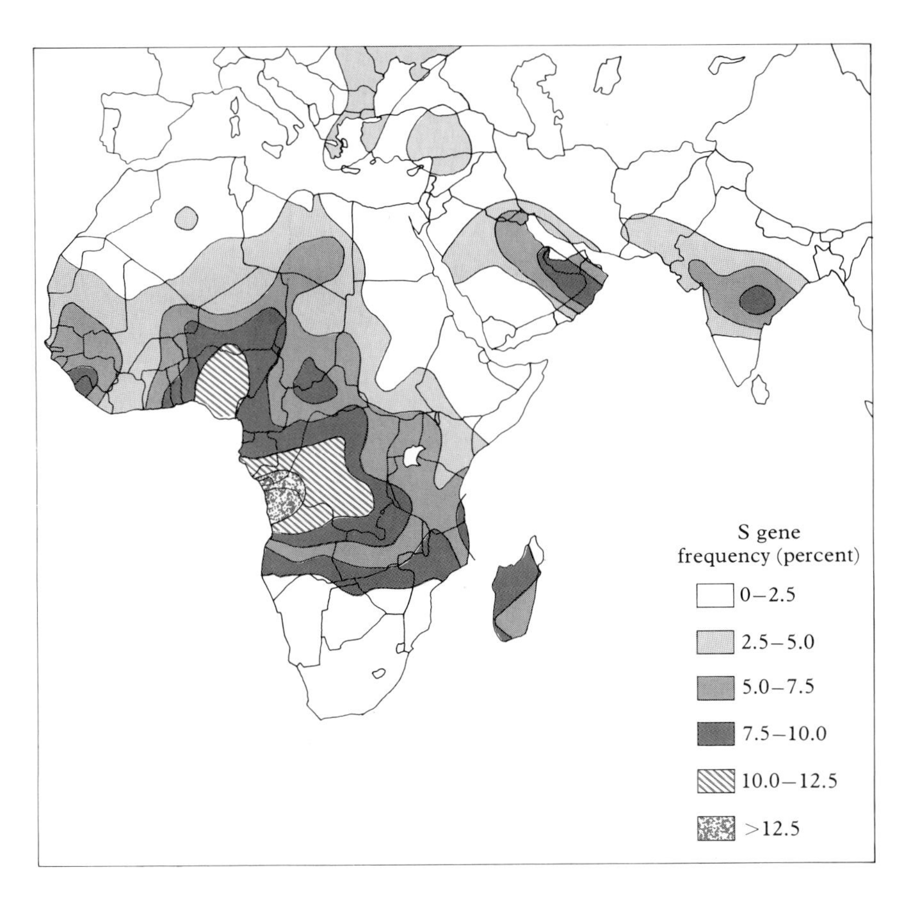

3–15
The distribution of the sickle-cell gene in various parts of the Old World, top, correlates with the distribution of malaria (dark areas), bottom, (From "The Genetics of Human Population," by L. L. Cavalli-Sforza. Copyright © 1974 by Scientific American, Inc. All rights reserved.)

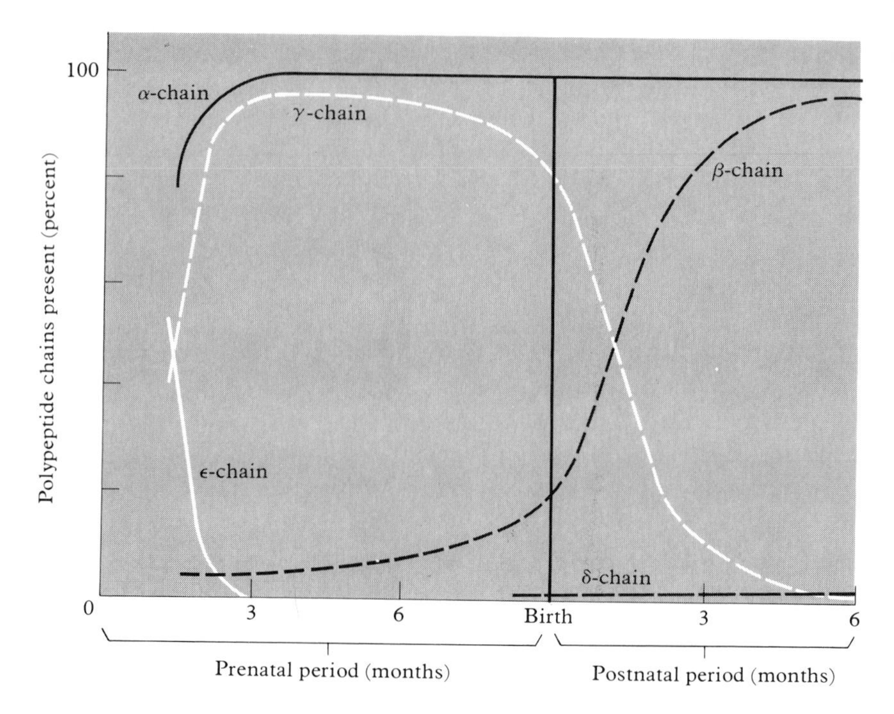

3–16
The percentages of some of the amino acid chains in normal hemoglobin molecules change markedly during development. (After Huehns, et al., "Human Embryonic Hemoglobins," Cold Spring Harbor Symposia on Quantitative Biology, 29, *1964.)*

As shown in Figure 3-16, the number of hemoglobin molecules containing two gamma, rather than two beta chains, changes dramatically shortly after birth. The number of molecules containing gamma chains falls off in proportion to the increase in the number of molecules containing beta chains. By the age of one year, gamma chains are normally found in only about 1 percent of hemoglobin molecules, and this percentage persists throughout adult life. Thus the gene coding for the gamma chain is very active before birth, but becomes almost inactive by the end of the first year of life.

Very early in development some of the hemoglobin molecules of fetuses contain still another chain known as the *epsilon chain.* The gene coding for the epsilon chain is active only during early embryonic life. As shown in Figure 3-16, hemoglobin molecules containing two alpha chains and two epsilon chains (known as *embryonic hemoglobin*) are present only during the first three months of development. Thus by the end of the third month, the gene coding for the epsilon chain is shut off and it never becomes active again.

Finally, shortly before birth the *delta chain* makes its first appearance. Hemoglobin molecules containing two alpha chains and two delta chains are called *Hemoglobin A_2*, and they are never very numerous. In normal adults Hemoglobin A_2 accounts for only about 2 percent of the total hemoglobin (see Figure 3-16). The components of normal human hemoglobins are summarized in Table 3-1.

The genes that code for all of these amino acid chains are examples of *structural genes,* so called because they bring about their effects by coding for the manufacture of specific protein molecules. But not all genes are structural genes. From the study of the genetics of bacteria we know that some genes, rather than coding for large protein molecules, have a controlling or *regulatory* role in protein synthesis. That this is probably also true of some human genes is suggested by certain genetic abnormalities of the kinds of hemoglobin we have just discussed.

In particular, some adults have been found to have abnormally high concentrations of fetal hemoglobin (two gamma chains). It has been suggested that a switch gene, or *regulatory gene,* may cause the rapid changeover from the synthesis of gamma chains to the synthesis of beta chains shortly after birth. Thus those persons who have abnormally high concentrations of fetal hemoglobin can be thought of as having a defective regulatory gene that failed to drastically slow the production of gamma chains.

Adults who have *only* fetal hemoglobin inside their red cells have been reported, and rather surprisingly this condition results in no apparent ill effects. This trait is autosomal recessive, and persons who are homozygous for it not only have no Hemoglobin A (two beta chains), but also lack the small amounts of Hemoglobin A_2 (two delta chains) that are normally present. It is known that the genes that code for the beta and delta chains are closely linked, which is to say that they are found close to one another on a strand of DNA. It has been suggested that persons whose red cells contain only fetal hemoglobin have a defect in the regulatory gene that normally controls the function of the two closely linked structural genes for the beta and gamma chains. One reason for favoring this explanation is that regulatory genes commonly control the function of two or more sequential structural genes in bacteria (see Figure 3-17). Overall, it seems that human regulatory genes do exist, but we have much to learn concerning how they bring about their effects.

From poorly understood regulatory genes we now turn our attention to some rather well-characterized structural genes that code for enzymes. Enzymes are protein molecules that speed the rate at which chemical reactions take place inside living cells, and thousands of them have been isolated from human cells. Human abnormalities resulting from defects in structural genes that code for enzymes are called *inborn errors of metabolism,* and they were first described in detail by the British physician Archibald Garrod in 1908 (Figure 3-18).

TABLE 3–1
Normal human hemoglobins.

CHAIN	COMBINES WITH TWO ALPHA CHAINS AND FOUR HEME GROUPS TO FORM	WHEN PRESENT	PERCENT OF TOTAL ADULT HEMOGLOBIN
Epsilon	Embryonic hemoglobin	Second and third month of fetal life	0
Gamma	Fetal hemoglobin	High fetal levels, drops to adult levels shortly after birth	1%
Beta	Hemoglobin A	From the second fetal month throughout life	97%
Delta	Hemoglobin A_2	From just before birth throughout life	2%

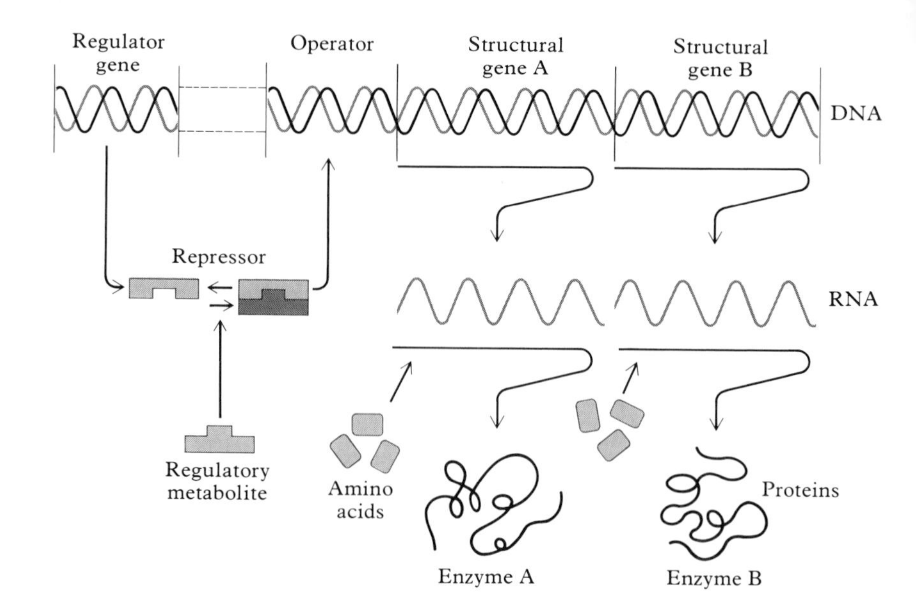

3–17
A model, based on bacterial genetics, of how regulatory genes control the synthesis of enzymes. The regulator gene directs the synthesis of a repressor substance that combines with a regulatory metabolite. This combination then influences the operator, which in turn shuts off the structural genes coding for the enzymes A and B.

Inborn Errors of Metabolism

3–18
Sir Archibald Garrod coined the phrase "inborn errors of metabolism" early in the twentieth century. (Courtesy of the Royal Society.)

One of the first abnormalities studied by Garrod was *alcaptonuria.* Persons who have alcaptonuria appear normal during childhood, but as adults they develop a blue-black discoloration of the ears, the whites of the eyes, the tip of the nose, and other areas of the body where cartilage lies just beneath the skin. Also, on exposure to several hours of sunlight the urine of those who have alcaptonuria turns jet black. Both the black urine and the discoloration of cartilage result from the buildup of a substance called *homogentisic acid.* This compound is also deposited in the cartilage of large joints, and those who have alcaptonuria may therefore develop severe arthritis.

In 1901 Garrod described 11 cases of this rare disorder and noted that at least three of the persons he studied were the offspring of parents who were rather closely related to one another, that is, three were the offspring of consanguineous matings. Garrod used this observation, along with his understanding of Mendel's ratios (which had been rediscovered only the year before), as the basis for a bold and insightful explanation of the nature of the defect that causes alcaptonuria. Garrod suggested that the condition was due to the effects of a single recessive gene that results in the manufacture of a defective enzyme. The disease was thus what Garrod called an "inborn error of metabolism."

Garrod's explanation proved to be correct. The exact nature of the biochemical defect in alcaptonuria is now known. As just mentioned, persons who have alcaptonuria have abnormally high concentrations of homogentisic acid in their urine and cartilage. The excess homogentisic acid results because they have very low, or nonexistent, concentrations of an enzyme that is usually responsible for the further metabolism of homogentisic acid.

Homogentisic acid is one of the metabolites of the amino acid *tyrosine.* In normal individuals the following metabolic pathway operates:

$$\text{Tyrosine} \xrightarrow{\text{enzyme}} \text{homogentisic acid} \xrightarrow{\text{enzymes}} \text{further metabolic products}$$

But in people who have alcaptonuria the lack of a specific enzyme results in a metabolic block to the further processing of homogentisic acid, which therefore accumulates in their tissues, where it may produce its undesirable effects. We can represent this block as follows:

enzyme enzymes
Tyrosine ——→ homogentisic acid —⟩⟨→ further metabolic products

Garrod later showed that *albinism,* a condition discussed in Chapter 1, is also the result of a metabolic block that affects the amino acid tyrosine. Tyrosine is not only the precursor of homogentisic acid, but also the basic building block of the pigment melanin.

enzymes → melanin
enzyme enzymes
Tyrosine ——→ homogentisic acid ——→ further metabolic products

The most common kind of albinism results from the lack of an enzyme that participates in the manufacture of melanin from tyrosine. (But albinism can also result because of a different metabolic defect that greatly reduces the amount of tyrosine available to form melanin. In this second kind of albinism affected persons have a normal concentration of the enzyme that facilitates the conversion of tyrosine to melanin.)

Since Garrod's time we have become aware of yet another inborn error of metabolism in the metabolic pathway we have been discussing. This defect is known as *phenylketonuria* (PKU). PKU is a serious disease in that if it is untreated it can result in severe mental retardation. PKU results because of a defect in the enzyme that converts the amino acid *phenylalanine* into tyrosine.

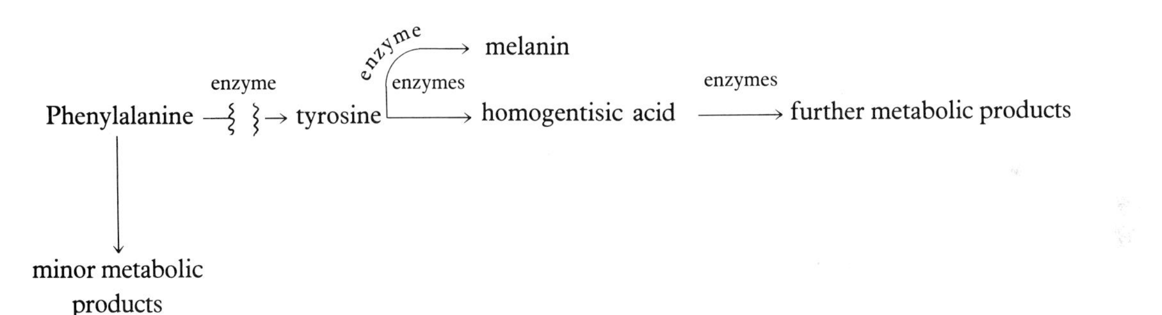

The lack of the appropriate enzyme for converting phenylalanine into tyrosine has two major effects. First, the concentration of phenylalanine in the tissues is greatly elevated. And second, minor metabolic products of phenylalanine that are normally present in only small quantities also accumulate. (It is probably the buildup of these minor metabolic products during the first few months and years of life that accounts for the mental retardation.) Because they cannot convert phenylalanine into tyrosine, persons affected by PKU are lightly pigmented. Nonetheless, enough tyrosine is directly available in the diet that they can still manufacture considerable quantities of melanin (and of homogentisic acid).

PKU can be effectively treated by greatly reducing the amount of phenylalanine present in the diet of an affected infant. Luckily for those who suffer from PKU, phenylalanine is one of the eight essential amino acids that the human body cannot manufacture for itself—all of the body's phenylalanine comes directly from the diet. By restricting dietary intake the buildup of high concentrations of phenylalanine and its minor metabolic products, and thus the mental retardation associated with the disease, can be prevented. At the same time, enough phenylalanine must be provided to allow for normal growth. Affected children are usually kept on their special diet until they are at least five years old, at which time the growth of the human brain is more or less completed.

Like most other inborn errors of metabolism, PKU is inherited as a recessive trait and is manifested only by homozygotes. But it is possible to identify carriers of the abnormal gene by subjecting them to the phenylalanine tolerance test. In this test a large dose of phenylalanine is administered orally and the rate at which it disappears from the bloodstream is measured. As shown in Figure 3-19, when heterozygous carriers of PKU are fed a standard dose of phenylalanine they show higher and more prolonged elevations of phenylalanine in their blood than do normal people. This is because heterozygous carriers of PKU, although they appear normal, have only about half of the normal concentration of the enzyme that aids in converting phenylalanine to tyrosine.

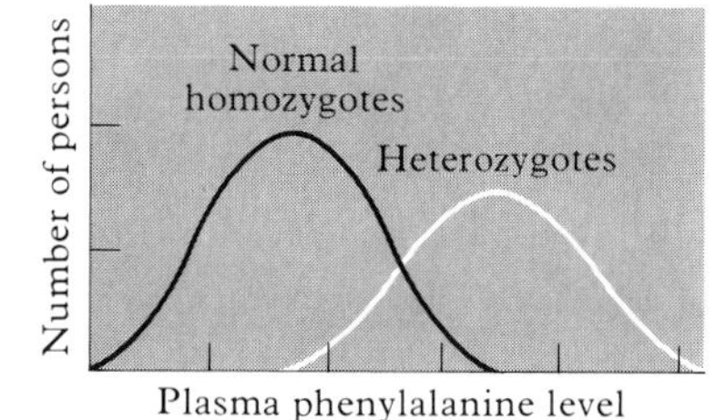

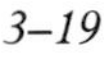

3–19
The phenylalanine tolerance test can identify heterozygous carriers of PKU.

In recent years the number of human diseases known to be due to inborn errors of metabolism has skyrocketed. Hundreds of diseases that result from abnormally low concentrations of critical enzymes have already been described, and the list continues to grow. But the ability to *treat* diseases caused by inborn errors of metabolism has not generally kept pace with the ability both to diagnose them and to identify normal-looking carriers by means of tolerance tests. It is sometimes possible to prevent the development of undesired effects by limiting dietary intake, as with phenylalanine and PKU. But more often than not we can do little or nothing to affect the course of diseases that result from inborn errors of metabolism. We are a very long way indeed from being able to correct the underlying defect in the genetic code of DNA that is ultimately responsible for the production of a faulty enzyme, or of no enzyme at all. Also, it is not now possible to treat these diseases simply by replacing the missing or defective enzymes. This is because enzymes generally do their work inside cells, and even if we are able to prepare concentrated extracts of the missing or defective enzyme, we have no way of getting them inside the cells that need them, where they speed up the rates of crucial biochemical reactions.

At the present time, and probably for some time to come, the best way to deal with diseases due to inborn errors of metabolism is to prevent them. There are two main ways of doing this. First, genetic counseling, sometimes in combination with tolerance tests, can help identify apparently normal carriers among the relatives of an affected person, or among other people who might be carriers. Second, it is now possible to detect many inborn errors of metabolism in cells obtained from human embryos during the first few months of development. Thus, parents might choose to voluntarily terminate a pregnancy if the fetus were proven to be affected by a serious inborn error of metabolism. (We will have more to say about how inborn errors of metabolism can be detected by prenatal diagnosis in Chapter 5.)

Summary

By the early 1950s experiments with bacteria and viruses had made it clear that the genetic material is not protein, but DNA. The Watson-Crick model of DNA structure was proposed in 1953. The DNA molecule consists of four kinds of nucleotides joined to one another in a double-stranded helix in such a way that the sequence of nucleotides in one strand automatically determines the sequence of nucleotides in the other strand. DNA replication is accomplished by the separation of the two strands; each strand then serves as a template for the manufacture of a new, complementary strand.

Biochemically, genes are specific regions along DNA molecules. Most genes have a role in protein synthesis. Proteins are made up of long chains of amino acids, and three successive nucleotides in the DNA molecule code for one amino acid in a particular protein. Protein synthesis requires a second nucleic acid, RNA, and the actual stringing together of amino acids into proteins takes place outside the nucleus on structures called ribosomes.

We have much to learn about how DNA is packaged inside eukaryotic chromosomes. We do know that some DNA segments are present in multiple copies and that DNA molecules are tightly coiled within the nucleus. Nuclear proteins provide a scaffold for DNA molecules and determine which genes are expressed when.

Persons affected by sickle-cell disease are homozygous for a gene that results in a single difference in the amino acid in the beta chain of their hemoglobin molecules compared with normal. But hemoglobin of persons who have sickle-cell trait (heterozygotes) is more resistant to malaria than is normal hemoglobin. And this probably accounts for the widespread distribution of the allele for the abnormal beta chain among people who live in areas where malaria is common—in spite of the reduced reproductive fitness of homozygotes.

Several different kinds of fetal hemoglobins have been described, and there is evidence that regulatory genes may bring about the rapid changeover in the composition of hemoglobin molecules that occurs during normal development. Regulatory genes commonly control two or more sequential structural genes in bacteria.

Garrod coined the phrase "inborn errors of metabolism" for human abnormalities resulting from defects in genes that code for enzymes. Alcaptonuria, albinism, and phenylketonuria are caused by defects in enzymes that speed up rates of metabolism of the amino acids phenylalanine and tyrosine. The best way to deal with inborn errors of metabolism is to prevent them. This can be accomplished by identifying carriers by means of pedigree analysis or tolerance tests, and identifying affected fetuses at early stages of development by means of prenatal diagnosis.

Suggested Readings

1. *Molecular Biology of the Gene,* 3d ed. by J. D. Watson. W. A. Benjamin, 1976. An authoritative overview of molecular genetics written to be understandable even to those who have little knowledge of chemistry.
2. "The Genetic Code: III," by F. H. C. Crick. *Scientific American,* Oct. 1966, Offprint 1052. A summary of how the sequence of base pairs in DNA provides information for the synthesis of specific protein molecules.
3. "The Visualization of Genes in Action," by O. L. Miller, Jr. *Scientific American,* March 1973, Offprint 1267. With the aid of the electron microscope one can "see" genes being transcribed into RNA and "watch" RNA being translated into protein.
4. "Chromosomal Proteins and Gene Regulation," by Gary S. Stein, Janet Swinehart Stein, and Lewis J. Kleinsmith. *Scientific American,* Feb. 1975, Offprint 1315. A discussion of the genetic role of histones and other proteins associated with chromosomal DNA.
5. "Repeated Segments of DNA," by Roy J. Britten and David E. Kohne. *Scientific American,* April 1970, Offprint 1173. The function of the repeated segments of DNA regularly found in eukaryotic cells has yet to be discovered.

This portrait of people of various human races is based on original photographs taken by Professor Carleton Coon, who kindly granted permission to have this drawing rendered. The groups represented are (right to left): front row, Armenian, Formosan, Bavarian, Veddoid; middle row, Dinaric, Singhalese, Arab; back row, Negrito, Korean, Swedish, Moroccan. *See also pp. 486–489.*

CHAPTER 4

Genes in the Human Population

A very distinguished group of British scientists, including Sir Ronald Fisher and Sir Julian Huxley, once visited the London Zoo to carry out a peculiar experiment. Their purpose was to find out whether the zoo's chimpanzees were capable of tasting a chemical known as phenylthiocarbamide (PTC). Among human beings, the ability to taste this unpleasantly bitter substance was known to be transmitted as an autosomal dominant trait. One of the scientist's main concerns was that they might not be able to tell whether individual chimpanzees could taste the chemical. But when Fisher gave a sip of a weak solution of PTC to the first chimpanzee to be tested, the creature's reaction left little room for doubt. The chimp was so disgusted by the taste that it spit in Fisher's face!

What prompted these talented scientists to perform such a seemingly foolish experiment? Some years earlier Fisher (Figure 4-1), most of whose formal education was in mathematics and statistics, had worked out the mathematical equations for describing how natural selection can influence the genetic composition of populations in which individuals mate with one another at random. In his book *The Genetical Theory of Natural Selection*, first published in 1930, Fisher showed that the diversity of genetically determined traits can be directly related to fitness for survival. In other words, Fisher's equations indicated that the genetic diversity observed in most populations exists in large part because a high degree of variability allows more possibilities for adapting to the environment and therefore for reproducing successfully. But at that time little was known of how extensive the variety of genetically determined traits is. By offering PTC to chimps at the London Zoo, Fisher and his distinguished colleagues discovered that chimpanzees, like their human relatives, vary in their abilities to taste the chemical PTC. They thus contributed (albeit little) to our knowledge of the range of variability among different individuals of the same species.

In recent years it has been discovered that individual animals of the same species are much more variable than Fisher or anyone else realized only a few decades ago. For the most part, our heightened awareness of how unique individual animals are has been the result of the study of genetically determined differences in protein molecules.

4–1
Sir Ronald Fisher was one of the first persons to formulate equations that describe how natural selection can influence a population's genetic composition. (Courtesy of Godfrey Argent.)

Genetic Variability in Protein Molecules

As discussed in the preceding chapter, protein molecules consist of long chains of amino acids linked in tandem by chemical bonds. There are about twenty kinds of amino acids, and the properties of a particular protein depend above all on the sequence of amino acids in its chain. What determines the sequence of amino acids in a protein is the sequence of the four bases along the particular stretch of the DNA molecule that codes for the protein in question.

A relatively easy and sensitive way to determine whether protein molecules differ from one another is to observe their patterns of movement in an electric field. Most protein molecules, or regions of protein molecules, have either a positive or a negative charge. The distribution of the charge depends on which amino acids are present in the protein and on how they are arranged. Proteins that differ from one another in their amino acid sequences usually have different electrical properties and therefore show different patterns of movement in an electric field. In general, these patterns reflect biochemical differences that are genetically determined. As you may recall, a difference of only one amino acid out of a total of 287 in the protein portion of the hemoglobin molecule can be easily detected by this method. (See Chapter 3.)

The variability of protein molecules, especially enzymes, has been the subject of intensive research in recent years. This systematic research has revealed a rather remarkable fact about the protein molecules of human beings and most other animals studied so far. At the biochemical level organisms of the same species are astoundingly diverse. This is because within a particular species, a given protein molecule is by no means identical in each individual.

Slightly different forms of almost all known human proteins have already been reported. Populations in which several alternate forms of a gene are regularly encountered are said to be *polymorphic* (many-formed) for that gene. As we have mentioned, the presence of some protein polymorphisms in the human population can be readily explained by natural selection. Thus the polymorphism of the beta chain of the hemoglobin molecule that results in sickle-cell disease in homozygotes is maintained in some geographic areas by natural selection because those who are heterozygous for normal and sickle-cell hemoglobin are more resistant to malaria than they would be if they had only normal hemoglobin inside their red cells.

But generally there is little or no evidence that a person is at an advantage or a disadvantage in having a slightly different form of a particular protein molecule. When the affected proteins are enzymes, it is usually possible to detect subtle differences in the activity of each form in the laboratory, but slight differences in protein molecules usually do not produce detectable ill or advantageous effects. Although it is possible to relate the persistence of some genetic polymorphisms in the human population to the effects of natural selection, this is generally *not* true at the biochemical level. Most often we have little or no evidence of how, or whether, natural selection maintains the extraordinary biochemical diversity of the human species. It may be true that many polymorphisms, especially those that do not produce obvious ill effects, are not affected by natural selection. For the present, we simply do not know.

In the rest of this chapter our main concern is with identifying some well-known polymorphisms in the human population and with explaining how the frequencies and distributions of these traits in the present-day population may have come about. We shall discuss the concept of *race* and attempt to explain how the frequencies of various genes observed in different segments of the human population can be affected, not only by natural selection, but also by isolation, chance, and other factors. Ultimately, all polymorphisms are the result of heritable changes in the sequence of the four bases in a particular region of DNA. Such changes in the DNA molecule are known as *mutations,* and in the pages that follow we will discuss the role of mutations in maintaining human diversity. But first, we turn our attention to a well-documented (though incompletely understood) genetic polymorphism about which reliable data are available from around the world. Perhaps the most extensively studied of any human polymorphism is the ABO blood group.

The Worldwide Frequencies and Distribution of the ABO Blood Group

The genes that determine to which ABO blood group an individual belongs come in three alternative forms, or alleles, that can be symbolized I^A, I^B, and I^O Each human being has one of the three alleles at the same location on each member of a particular pair of autosomes. The patterns of inheritance shown by these three alleles and how they determine to which ABO blood group a person belongs are discussed in Chapter 1. The worldwide frequencies of these three alleles in the human population are known with considerable accuracy, and, as determined by medical and anthropological surveys, the worldwide frequencies are: I^O, 62 percent; I^A, 22 percent; and I^B, 16 percent.

But these alleles are not distributed equally throughout the world. Figure 4-2 shows the distribution of the allele I^O in aboriginal populations around the world. Notice that North and South American Indians have very high frequencies of the allele I^O. In fact, in some areas of these two continents, the aboriginal peoples have the allele I^O almost to the exclusion of the other two. On the other hand, the distribution of the allele I^B is almost opposite to that of I^O, as shown in Figure 4-3. I^B is very common in central Asia, but rare among native North and South Americans.

How can we account for these definite patterns in the worldwide distribution of the alleles that determine to which ABO blood group a person belongs? Perhaps the most likely explanation is that the distribution of ABO blood groups is one effect of human migrations, most of which took place in prehistoric times. Human beings as we know them today probably evolved first in Africa or Asia.

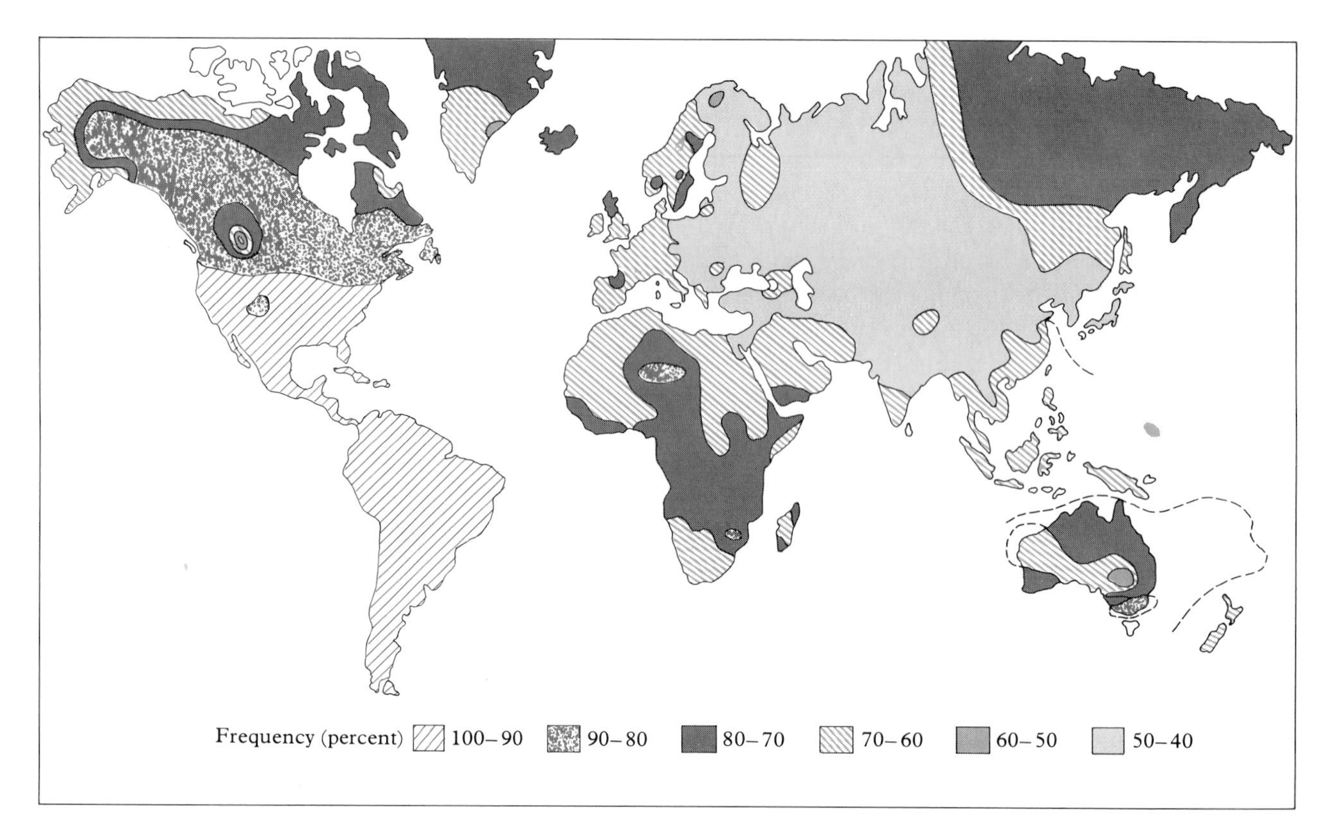

4–2
The distribution of the allele I^{O} *in aboriginal populations around the world. (After Mourant et al.,* The ABO Blood Groups. *Copyright © 1958 by Blackwell Scientific Publications.)*

Initially human populations must have been rather small, but with their characteristic resourcefulness people soon invented food raising and then increased their numbers and extended their range, not only to the entire African and Asian continents, but to the rest of the Old World as well. Eventually, people made their ways to the Americas, the islands of the Pacific, and Australia. These migrations could easily result in the gradual emergence of the present-day distribution of the ABO blood groups.

One example of how the distribution of ABO blood groups can be influenced by human migrations is shown in Figure 4-4, which is a close-up map of the distribution of the allele I^{B} in Eurasia. In general, the frequency of I^{B} shows a steady decrease from Central Asia toward Western Europe. This pattern has been attributed to the effects of Mongol invasions of Europe, which continued for about 1,000 years and ended about 500 years ago. The Mongols from the East may have had proportionately higher concentrations of I^{B} than their Western counterparts, and as the former moved westward they undoubtedly spread not only their culture but their genes as well. The distribution of Group B in Western Europe could be explained by assuming that before the Mongol invasions most West Europeans did not have the allele I^{B}. A comparison of Figures 4-2 and 4-4 suggests that early West Europeans probably had a preponderance of the allele I^{O} instead, as evidenced by the higher frequency of I^{O} in extreme Western Europe today. (The very low concentrations of I^{B} that are found today in the area of the Pyrenees and Caucasus mountains may have resulted because some of the original Europeans fled to the mountains to evade the Mongol hordes.)

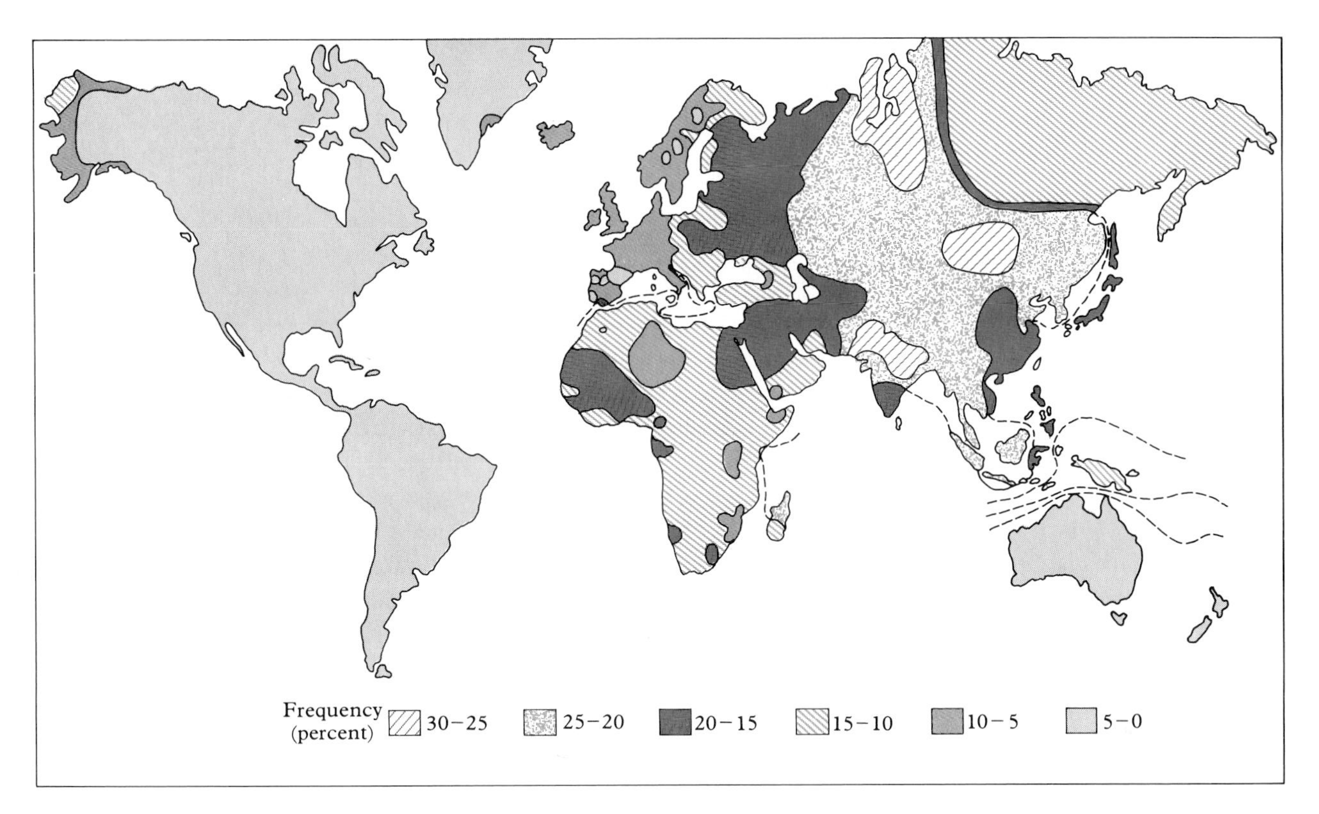

4–3
The distribution of the ellele I^B *in aboriginal populations around the world. (After Mourant et al.,* The ABO Blood Groups. *Copyright © 1958 by Blackwell Scientific Publications.)*

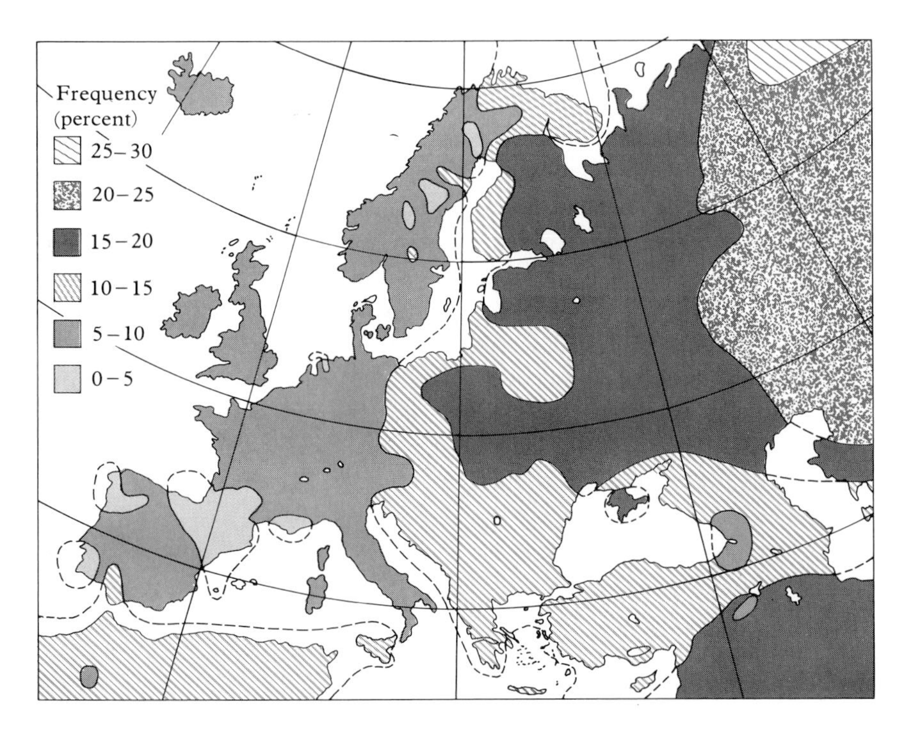

4–4
The distribution of the allele I^B *in Eurasia. Note the relatively low frequencies of* I^B *near the Pyrenees and Caucasus mountains. (After Mourant et al.,* The ABO Blood Groups. *Copyright © 1958 by Blackwell Scientific Publications.)*

Overall, there is little doubt that human migrations have strongly influenced the present distribution of the ABO blood groups. But what about the effects of natural selection? The difficulty in trying to assess its importance in determining the frequencies and distribution of the ABO blood group is that we have no decisive evidence about whether any one of the three alleles is directly influenced by natural selection or not. That is, human beings of one ABO blood group enjoy no obvious reproductive advantage over those of any other ABO group.

But this is not to say that statistical correlations between ABO blood groups and various diseases do not exist. In fact, based on data from Great Britain in the 1950s there is evidence that people of Group O are about 40 percent more likely to develop a duodenal ulcer than members of the other two blood groups. What is not clear, however, is how or whether duodenal ulcers are related to natural selection. If the two are related, then we would expect people who have ulcers to produce either more or fewer offspring than people who do not, but there is no evidence that this happens. It seems unlikely that natural selection has influenced the worldwide distribution and frequencies of the ABO blood groups simply because of the statistical correlation between Group O and ulcers. Besides, the various ABO groups have been statistically correlated, not only with duodenal ulcers, but with many other traits, such as cancer of the stomach and the hard-to-define characteristic of "tender-mindedness" versus "tough-mindedness."

On the other hand, it has been suggested that natural selection may have influenced the present-day distribution and frequencies of the ABO blood groups because people of different groups may be more resistant to certain infectious diseases. For example, people of Groups B and O have natural antibodies against Group A in their bloodstreams. Antibody against Group A may also be effective against the virus that causes smallpox. Thus, members of Groups B and O may be more resistant to smallpox than those of Group A, and the present-day frequencies and distribution of the ABO groups could reflect devastating epidemics of smallpox or other infectious diseases in the past.

Natural selection could also affect the ABO groups because of antibody-dependent incompatibilities between a mother and her developing fetus. For example, if a woman is Type O and her fetus is Type A, there may be difficulty because the mother has natural antibodies against the red cells of the fetus. But data about the actual outcome of ABO-incompatible pregnancies—although abundant—are inconclusive and some of them are contradictory. For now, we don't know whether natural selection affects the frequencies of ABO blood groups because of antibody-dependent incompatibilities between mother and fetus.

To sum up, natural selection has undoubtedly played a role in determining the frequencies and distribution of the ABO blood groups, but it has probably not been the most important factor. Rather, the present-day frequencies and distribution of the ABO blood groups in the human population can perhaps best be explained by prehistoric migrations and the effects of chance. That chance can play a decisive role in determining the genetic composition of certain human populations is well known. One example of how this occurs is provided by the Dunkers, a group of people who migrated from Germany to Pennsylvania in the early eighteenth century.

The Dunkers and Genetic Drift

In populations made up of large numbers of people who mate with one another at random, the frequencies of those genes not obviously influenced by natural selection tend to remain about the same from one generation to the next. This is because the frequencies of such genes are determined mainly by chance, and in large populations chance fluctuations in gene frequencies tend to balance one another. But in relatively small populations this does not necessarily happen, and significant changes in the genetic composition of the population can occur by chance alone. This chance variation in gene frequencies from one generation to another is known as *genetic drift,* and in general the smaller the population the greater the genetic drift can be.

An ideal population in which to examine the effects of genetic drift is the devout Protestant religious sect known as Dunkers (Figure 4-5). Between 1719 and 1729 fifty families of Dunkers emigrated from the German Rhineland to Pennsylvania and thereby completely transplanted the sect to the New World.

4–5
The Old Order Dunkers of Franklin County, Pennsylvania. Although they seldom marry outside the sect and their attire differs from that of their neighbors, their customs are not otherwise unusual. (From "The Genetics of the Dunkers," by H. Bentley Glass. Copyright © 1953 by Scientific American, Inc. All rights reserved.)

To marry outside the church is considered a grave offense, and a Dunker who does so must either withdraw voluntarily from the community or be expelled from it.

During their first hundred years in Pennsylvania the Dunkers doubled in number, and almost all of them could trace their ancestry to the original fifty families. Then in 1882 the Dunker church underwent schism and a progressive group separated from the old order. At that time the Old Order Dunkers numbered about 3,000, and this number has not changed much to the present day. One of the original Dunker communities, in Franklin County, Pennsylvania, remained with the old order, and in the years since 1882 its size has also changed very little. When the Franklin County group of Dunkers was studied in the early 1950s, the population was about 300 and its size had been nearly the same for several generations. The Franklin County Dunkers were thus an almost ideal population in which to look for the effects of genetic drift.

The effects of genetic drift on the Franklin County Dunkers can be revealed by a comparison of the frequencies of various genes among the Dunkers, their West German forebears, and their present-day American neighbors. For example, these are the frequencies of the ABO blood group alleles I^O, I^A, and I^B among the Dunkers, West Germans, and Americans:

	I^O	I^A	I^B
Dunkers	60%	38%	2%
West Germans	64%	29%	7%
Americans	70%	26%	4%

Notice that the frequencies of these alleles among the Dunkers are not the same as those of West Germans or Americans; nor do they lie between the two. This is to be expected if genetic drift is at work and the frequencies of the alleles are determined largely by chance.

Blood groups are not the only characteristic of the Dunkers that shows the effects of genetic drift. There are clear-cut differences between the Dunkers and surrounding American communities in the frequencies of several other apparently nonadaptive traits. Thus, as compared with their neighbors, fewer Dunkers have hair on the middle segment of one or more fingers, fewer are able to bend the end of the thumb backwards to form an angle of more than 50°, and fewer have earlobes attached to the side of their heads rather than hanging free (Figure 4-6). The best explanation for these observations is that the frequencies among the Dunkers are the result of genetic drift.

The Dunkers are not the only human population in which genetic drift has been detected. Marked variations in the frequencies of certain alleles from those of neighboring populations have also been reported in a number of populations, including aboriginal Australians, Eskimos, Italians in isolated villages, North American Indians, and religious sects in Montana. All of these groups have in common two important features that enable genetic drift to strongly influence their genetic constitutions. First, the populations are relatively small, and second, the groups are isolated from neighboring populations either by physical barriers or by cultural rules.

Although the effects of genetic drift are most pronounced in the smallest populations, genetic drift can also influence the genetic composition of larger populations. It has been estimated that genetic drift can strongly influence the

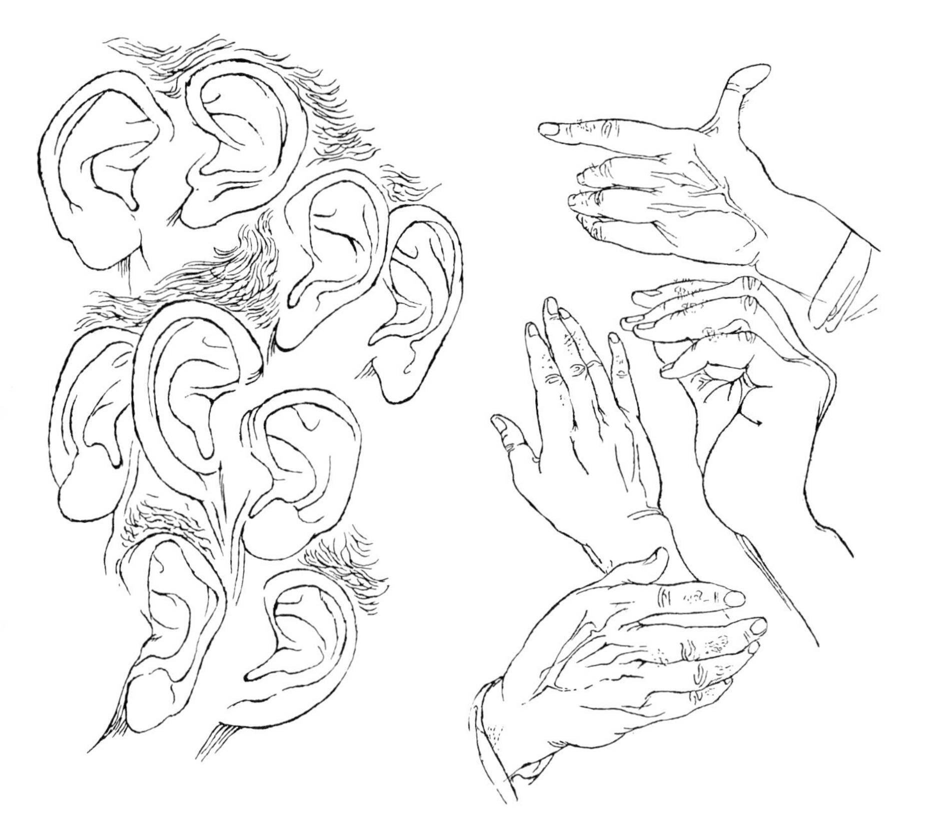

4–6
Some characteristics that show the effects of genetic drift in the Dunkers. Earlobes can be attached or they can hang free; those of most Dunkers hang free. Dunkers have a high incidence of hitchhiker's thumb (the ability to bend the thumb backwards at an angle of more than 50 degrees); and, as compared with the general population, fewer Dunkers have hair on the middle segment of one or more fingers. (From "The Genetics of the Dunkers," by H. Bentley Glass. Copyright © 1953 by Scientific American, Inc. All rights reserved.)

frequencies of apparently nonadaptive traits in human populations if the parents in any generation number a few hundred individuals or fewer. This is important because before the invention of food raising 10,000 years ago most human populations were probably within this size range and were therefore small enough to be strongly influenced by genetic drift. In fact, many inherited differences between human beings belonging to different races may have become established thousands of years ago when people lived in small groups that were physically and reproductively isolated from one another. Let us discuss the concept of *biological race* and its relation to genetic drift and to natural selection.

The Biology of Human Races

The species to which all living people belong, known as *Homo sapiens,* originated in either Africa or Asia and then spread outward. From the start our species has been endowed with dextrous, toolmaking hands that are controlled by the most complicated organ that evolution has produced so far—the human brain. The combination of human hand and human brain can be thought of as an adaptation that has allowed our species to virtually cover the surface of the land. No other species is as widely distributed as the human species, with the possible exception of species such as houseflies, body lice, and mice, which directly benefit from human activities and which have therefore followed people in their migrations over the continents.

When widespread species are examined over their full geographic range, it is often found that populations in different places look slightly different. For example, song sparrows from New York and Oregon can easily be told apart, as can zebras from different parts of Africa (Figure 4-7). As we all know, the human species is no exception when it comes to geographic variation. Thus, people from Tokyo, Copenhagen, Bombay, and Nairobi can be told apart as easily as the song sparrow and zebras from different locations. What accounts for the differences in appearance between populations of the same species?

These variations are the result of the interaction of the genes of a population with the environment. The environment determines which genes from among the total range of genes in the species' DNA will be expressed and to what degree they will be expressed. Nonetheless, most populations of widespread species look slightly different from one another because of genetic differences between the groups. These genetic differences arise because populations that are separated from one another by long distances or other barriers cannot interbreed. Whenever local populations of a given species are isolated from one another by a barrier, genetic differences accumulate because of the effects of natural selection, mutation, and, if the populations are small enough, genetic drift. Distinct local breeding groups may thus arise in different areas of a widespread species' range. These distinct local breeding groups within a particular species are known as *races.*

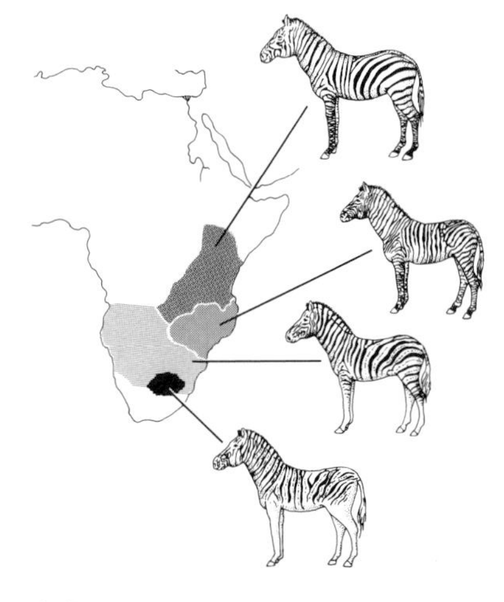

4–7
Variation of stripes of African zebras in different geographic regions. (After Cabrera, Journal of Mammology 17, *1936.)*

There is no doubt that human races originated in the same way as other animal races. About 40,000 years ago most human beings probably lived in small tribes that were relatively isolated from one another by distance and custom. This isolation resulted in chance differences in genetic constitution from one group to another. During this time of relative isolation among early human groups some of the genetic traits that characterize modern races probably became established by genetic drift.

But chance was not the only factor favoring the development of genetic differences between isolated groups of ancient people. Surely natural selection must also have played a role in the evolution of human races. As our ancestors increased in number and extended their range, different groups found themselves in very different environments. You will recall that because of natural selection organisms tend to acquire traits that allow them to closely adapt to the local environment. Thus it seems likely that at least some racial characteristics became established because they were advantageous under certain environmental conditions.

Perhaps the most apparent human racial characteristic is the color of a person's skin. At least 36 shades of human skin color, ranging from jet black to almost white, have been described. As shown in Figure 4-8, there is undoubtedly a worldwide connection between skin color and the intensity of sunlight (especially its ultraviolet component). In general, the most darkly pigmented people live closest to the equator and are exposed to the greatest concentration of sunlight. In some parts of the world, especially Africa, skin color shows a steady gradation from darker to lighter the further away from the equator the population lives. In other locations changes in pigment with latitude are not so clear cut, and the original pattern has been blurred and distorted by human mobility and racial mixing in recent years.

It has been suggested that natural selection influences the worldwide distribution of human skin color because the amount of pigmentation is related to the synthesis of Vitamin D. Vitamin D plays a rather unusual role in human nutrition for several reasons. First, Vitamin D does not exist naturally except in the livers of certain kinds of animals. Instead, some foods contain precursors

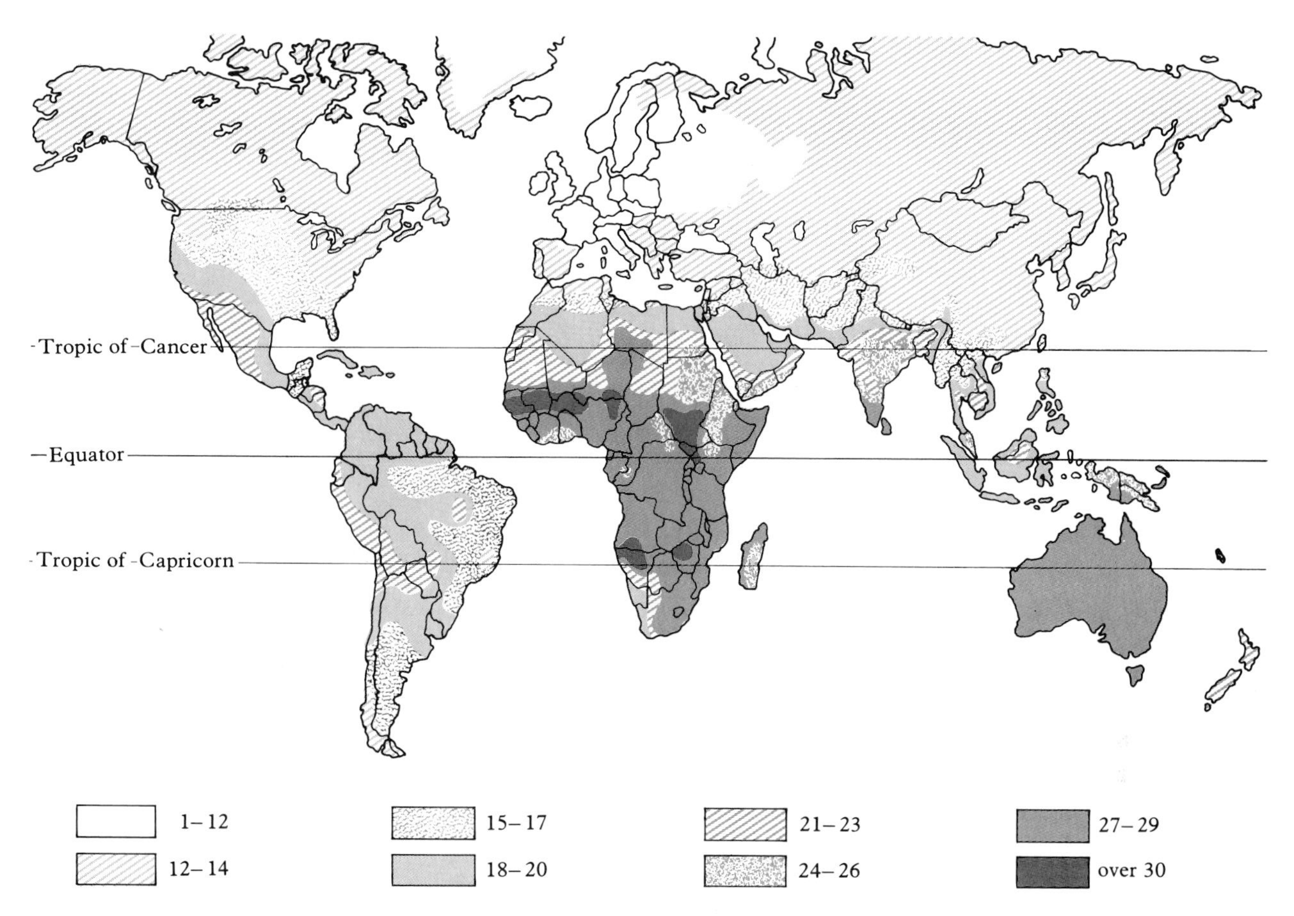

4–8
The distribution of human skin color before Columbus's first voyage to the New World in 1492 A.D. The values increase with darker skin color. (After R. Baisutti, Razze e Popoli della Terra, *Torino: UTET, 1951.)*

of Vitamin D, and from these precursors Vitamin D is synthesized by the human skin as long as energy from ultraviolet radiation (sunlight) is available. Second, unlike most human vitamins, Vitamin D is toxic in large doses. Vitamin D in normal quantities is essential for proper calcium metabolism, but in excess it can cause kidney stones and can lead to the production of deposits of calcium in areas of the body that are not usually calcified. (For example, calcium deposits may build up in the walls of large arteries or in the cornea of the eye.)

How does the worldwide distribution of human skin color relate to Vitamin D? A possible, though speculative, explanation is as follows. People who live closest to the equator receive the most sunlight and their skins therefore synthesize much more Vitamin D than those of people who live further away from the equator. People near the equator generally have darker skins because the accumulation of pigment in the deeper layers of skin absorbs untraviolet radiation and thus prevents the synthesis of excess Vitamin D. On the other hand, people in northerly latitudes, where ultraviolet radiation is much less intense, tend to have lighter skin so that their bodies can use the available radiation to synthesize sufficient Vitamin D to maintain good health. (Too much Vitamin D does not accumulate in light-skinned people during summertime because their skins become tanned and thus screen out the seasonal excess of ultraviolet radiation. In fact, the tanning reaction is initiated by light of the same wavelength needed for Vitamin D synthesis.)

Because our species almost certainly evolved first in the tropics, it is likely that all human beings were at first darkly pigmented. During the early stages of human evolution natural selection may have favored dark skin, not only because of the relation between Vitamin D and pigmentation, but also because a dark color may have provided better camouflage for our ancestors, who must have been prey for some of the large carnivores of the day. As people migrated from the tropics to more northerly or southerly regions they probably lost most of their skin pigmentation because in doing so they were able to synthesize sufficient quantities of Vitamin D. (Interestingly, the Eskimos are the only northerly people who have dark skins, and they have abundant supplies of Vitamin D in their diets because they regularly eat fish livers.)

You may be wondering why people possess some genetically determined traits, such as skin color, in seemingly endless varieties that differ slightly whereas they either have or do not have others, such as the ability to taste the chemical PTC. The main reason is that the former are under the influence of several pairs of alleles, whereas the latter depend on a single pair. Thus the regular variation in skin color among human populations around the world is the result of the interaction of at least four alleles, as shown in Figure 4-9. That

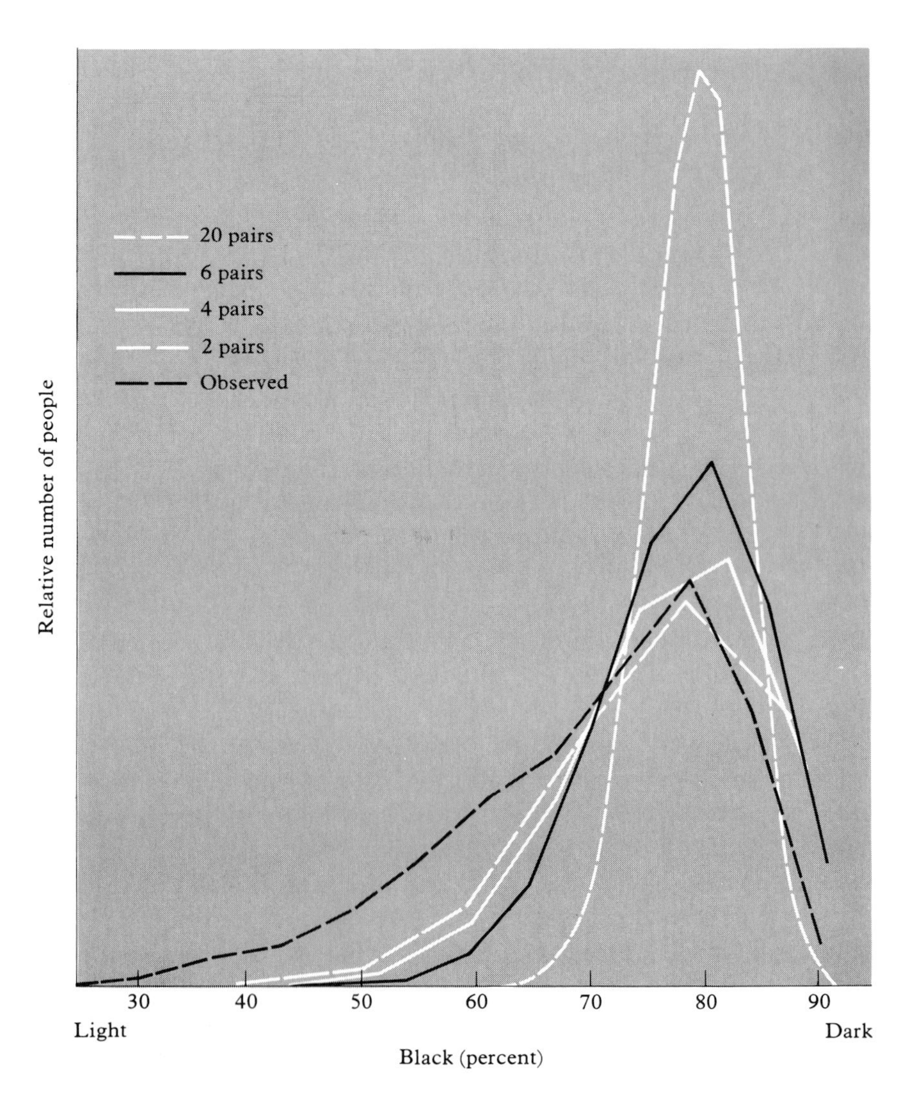

4–9
The distribution of the observed skin color of American blacks and the distribution that would be expected if two, four, six, and 20 pairs of alleles were acting together. (After Curt Stern.)

these alleles are (or were) retained in different frequencies in different human populations can probably be attributed to the effects of natural selection.

Another characteristic that can show gradual variation with latitude, that depends on several pairs of alleles, and that is influenced by natural selection is body build. In general, warm-blooded animals that live in hot equatorial climates are smaller, have longer arms and legs, and have larger ratios of surface area to body weight than warm-blooded animals that live farther north or south. Most of these tendencies are not very clear-cut in the human population, and there are a lot of exceptions to the rule. Nonetheless, as shown in Figure 4-10, human adaptation to climate probably accounts for the differences in body build that allow tall, slender Africans to dissipate more unneeded body heat than Eskimos, who because of where they live must conserve as much body heat as possible.

It is likely that other body features, such as the size of the nose, the color and form of the hair, and the shape of the eyefolds also became established in certain populations because of natural selection. This is because all of these characteristics are external, as are skin color and body build. We would expect the superficial, readily visible characteristics to experience the effects of natural selection because our body surfaces are the interface between our bodies and the environment. It is therefore no wonder that human body surfaces have been altered by natural selection to best fit the varied environments into which our ancestors migrated (Figure 4-11).

Biochemical Differences Between Human Races

But differences in traits that are neither superficial nor readily visible parallel the superficial differences of human races to only a limited degree. You will recall that in all species protein molecules, especially enzymes, usually exist in several alternate forms. In recent years it has been possible to compare differences in protein molecules within and between the various races of people. For those protein polymorphisms not obviously influenced by natural selection (and this includes the great majority), the differences in protein molecules between living human races are slight. And protein differences are as slight between Caucasians and African blacks as between Caucasians and Orientals. Moreover, the protein molecules of two Caucasians from opposite ends of Europe differ more than the molecules of two Caucasians from the same isolated European village, but in both differences are about the same as those between the molecules of Caucasians and African blacks or Orientals. Thus at the biochemical level, the differences between human races are generally much less pronounced than the superficial differences that are apparent on body surfaces. In fact, there is so much biochemical overlap that people cannot be assigned with certainty to a given race solely on the particular alleles contained with their genetic programs.

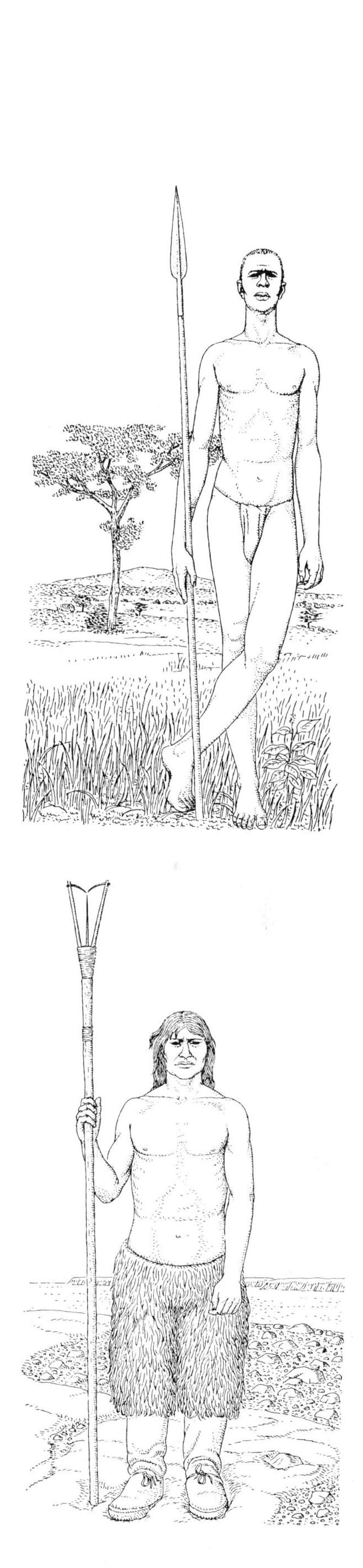

4–10

The greater body surface of the Nilotic Negro from the Sudan (top) allows him to dissipate unneeded body heat, whereas the proportionately greater bulk of the Eskimo (bottom) conserves body heat. (From "The Distribution of Man", by William W. Howells. Copyright © 1960 by Scientific American, Inc. All rights reserved.)

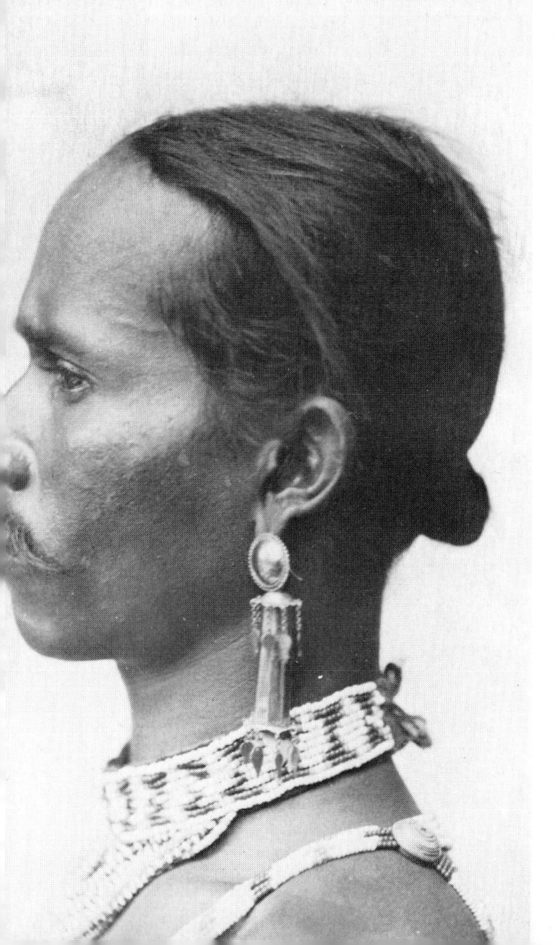

4–11
As shown in these photos, members of the human species are wonderfully diverse in the contours of their body surfaces. The superficial differences between human races probably resulted from the effects of natural selection on our distant ancestors in various parts of the world. Opposite page, top left, Kalahari bushman with two handfuls of stork; top right, young Negrito woman and her children; bottom, Blackheart, a Blackfoot Indian. This page, top, a young Polynesian woman, probably Tahitian; bottom, a Singhalese man (continued on the next two pages).

4-11 (continued)
Opposite page, top left, Melanesian man from Admiralty Islands; top right, Singhalese woman dancer; bottom, young woman from Inner Mongolia. This page, top, three young Swedish women; bottom, Mamayauk Eskimo woman. (Photographs on pp. 486–489 courtesy of American Museum of Natural History.)

The Number of Human Races

How many human races are there? That depends on how the term is defined. The best definition is that races are local breeding groups within a particular species. Races are thus defined by genetic relations between populations and not by differences between individuals. But most local breeding groups in any species tend to blend with one another at their geographic boundaries because adjacent races interchange genes when they reproduce. Thus, races are never clear-cut, precisely defined entities. There are undoubtedly thousands of local breeding groups within the human population today that could legitimately be defined as races. The Dunkers are one example, as are relatively isolated populations in Alpine villages, New Guinea, and Australia.

But it is neither practical nor particularly informative to designate every isolated group of people as a separate race. Rather, most of the time the term *race* is used for any relatively isolated local breeding group that is convenient to distinguish for purposes of a given study. In other words, the number of human races described depends in large part on the purposes of the describer.

For the most part, the description of human races has been undertaken, appropriately, by anthropologists, but some of the criteria used to distinguish individual races are at best vague. The number of human races currently recognized by different anthropologists using different criteria ranges from zero up, but most estimates fall somewhere in between three and thirty human races (see Table 4-1). The criteria used to distinguish human races include not only features of the body surface, but also less obvious physiological and biochemical differences. In general, the genetic differences between currently recognized human races reflect some degree of reproductive isolation, but there is no denying that because of technologic and sociologic changes human populations are much less isolated from one another than ever before. Although human races do exist, it is questionable whether they have much biologic relevance for our species at the present time. Let us discuss this important point further.

The Importance of Culture in Human Adaptation

Among the million and a half species of living things that have been described and named, the human species is unique in that its members adapt to the environment primarily by means of a complicated form of learned behavior called *culture,* which is transmitted from generation to generation by means of the symbol system of language. For all practical purposes geographic variation in our species is now irrelevant because people adapt to the environment primarily by means of behavior, whose biological basis is in the brain and is not reflected in superficial differences in body surfaces. When they first evolved, people who had dark-colored skins were better adapted to climatic conditions near the equator than those who had light-colored skins, but few human beings now live under "natural" conditions. Also, technological advances have assured that at the present time human beings of various races are as likely to reproduce in one environment as another. The superficial differences between existing

TABLE 4–1

Various human "races" that can be identified by means of statistical correlations of the structures of a variety of protein molecules. This analysis was made by the population geneticist R. C. Lewontin and was based on data about 17 proteins. Seven races and many distinct populations can be recognized.

Caucasians
Arabs, Armenians, Austrians, Basques, Belgians, Bulgarians, Czechs, Danes, Dutch, Egyptians, English, Estonians, Finns, French, Georgians, Germans, Greeks, Gypsies, Hungarians, Icelanders, Indians (Hindi speaking), Italians, Irani, Norwegians, Oriental Jews, Pakistani (Urdu-speakers), Poles, Portugese, Russians, Spaniards, Swedes, Swiss, Syrians, Tristan de Cunhans, Welsh

Black Africans
Abyssinians (Amharas), Bantu, Barundi, Batutsi, Bushmen, Congolese, Ewe, Fulani, Gambians, Ghanaians, Hobe, Hottentot, Hututu, Ibo, Iraqi, Kenyans, Kikuyu, Liberians, Luo, Madagascans, Mozambiquans, Msutu, Nigerians, Pygmies, Sengalese, Shona, Somalis, Sudanese, Tanganyikans, Tutsi, Ugandans, U.S. Blacks, "West Africans," Xosa, Zulu

Mongoloids
Ainu, Bhutanese, Bogobos, Bruneians, Buriats, Chinese, Dyaks, Filipinos, Ghashgai, Indonesians, Japanese, Javanese, Kirghiz, Koreans, Lapps, Malayans, Senoy, Siamese, Taiwanese, Tatars, Thais, Turks

South Asian Aborigines
Andamanese, Badagas, Chenchu, Irula, Marathas, Naiars, Oraons, Onge, Tamils, Todas

Amerinds
Alacaluf, Aleuts, Apache, Atacameños, "Athabascans," Ayamara, Bororo, Blackfeet, Bloods, "Brazilian Indians," Chippewa, Caingang, Choco, Caushatta, Cuna, Diegueños, Eskimo, Flathead, Huasteco, Huichol, Ica, Kwakiutl, Labradors, Lacandon, Mapuche, Maya, "Mexican Indians," Navaho, Nez Percé, Paez, Pehuenches, Pueblo, Quechua, Seminole, Shoshone, Toba, Utes, "Venezuelan Indians," Zavante, Yanomama

Oceanians
Admiralty Islanders, Caroline Islanders, Easter Islanders, Ellice Islanders, Fijians, Gilbertese, Guamians, Hawaiians, Kapingas, Maori, Marshallese, Melanauans, "Melanesians," "Micronesians," New Britons, New Caledonians, New Hebrideans, Palauans, Papuans, "Polynesians," Saipanese, Samoans, Solomon Islanders, Tongians, Trukese, Yapese

Australian Aborigines

Source: From R. C. Lewontin in *Evolutionary Biology*, Vol. 6. T. Dobzhansky et al. (eds.). Plenum Publishing Corp. 1972.

human races are thus relics of the past in large part and are not of much functional significance today.

In the evolutionary sense, the significance of races is that under special circumstances some may evolve into new species. The term *subspecies* is used to describe a race that is sufficiently distinct to merit (in the opinion of the person who is classifying) a Latin name in a formal classification. Given enough time and the presence of significant changes in the environment, some isolated groups may become so genetically different from other groups that they can no longer reproduce with one another successfully. When that happens, a new species may evolve.

The amount of variation observed between populations of living human beings probably does not warrant classifying any of them as official subspecies, especially because racial differences have been blurred as recent years have brought people increased mobility and social changes. Nor is there any evidence that groups of people are becoming reproductively isolated from one another and thus may be in the early stages of evolving into a new species. Even the recently abolished castes of India, which were reproductively isolated from one another for at least 3000 years, showed no signs whatsoever that individuals of particular castes were incapable of successfully reproducing with members of any other caste or a member of any other human population.

But subspecies are officially recognized in extinct members of the human species. Thus Neanderthalers are officially classified as a distinct subspecies, *Homo sapiens neanderthalensis.* This emphasizes that some Neanderthalers probably interbred with fully modern people, known as *Homo sapiens sapiens.* In fact, modern people may well have evolved directly from isolated groups of Progressive Neanderthalers.

In the end, the wondrous diversity of living things, including our own species, has probably arisen in large part because of the accumulation of genetic differences between isolated populations. The original source of genetic differences between two individuals is found in the phenomenon known as mutation. We now turn our attention to how heritable changes originate within the genetic material and thus produce genetic diversity in any population.

Mutations and Human Diversity

All of the differences between any two living things ultimately originate because of changes that take place within DNA molecules. Heritable changes that arise within existing DNA molecules are known as *mutations,* and in general mutations are the results of accidents. One of the main ways in which mutations arise is by mistakes that occur during DNA replication. As you know, the sequence of the four bases in one strand of a DNA molecule is complementary to the base sequence of the other strand. Thus, when the two strands separate during DNA replication, each strand provides the base sequence necessary to code for a new complementary partner and the two molecules that result are identical, as long as errors in the copying process do not occur. Most of the time the copying process is completely accurate, but rare mistakes do happen, and these unlikely errors help to furnish the raw material for evolution.

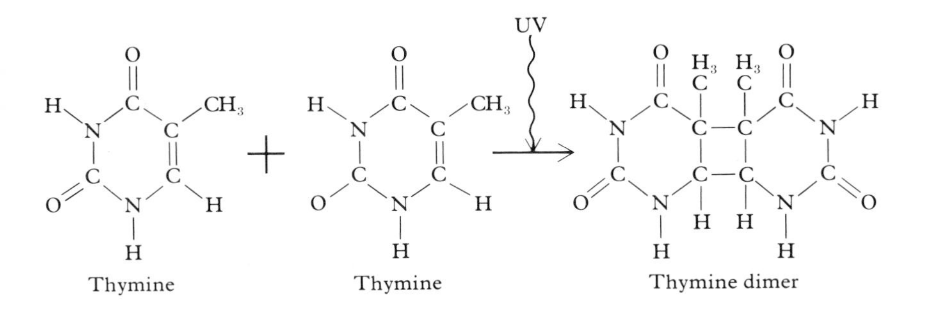

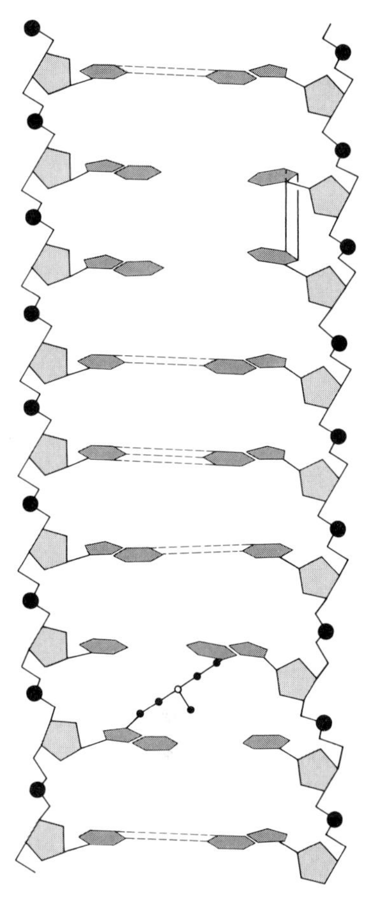

4–12
Thymine dimers can be formed when ultraviolet light interacts with thymine bases in DNA molecules, left. Thymine dimers distort DNA molecules so that they cannot replicate properly, right. (Right-hand portion from "The Repair of DNA," by Philip C. Hanawalt and Robert H. Haynes. Copyright © 1967 by Scientific American, Inc. All rights reserved.)

Several sorts of errors can occur during DNA replication. In brief, most replication errors are caused by the substitution of one base pair for another. This may occur because of the accidental mismatching of two bases that do not usually pair with one another or because of the insertion or excision of one or several base pairs along a particular stretch of DNA. (But there are many other reasons.) In any case, the end result is that a portion of the genetic code is altered. If the alteration in the genetic code occurs in a structural gene (that is, in a stretch of DNA that codes for a particular protein molecule), then we would expect the protein whose amino acid sequence is coded for by the altered gene to be altered too. It turns out that most alterations in protein structure by the kinds of mutations we have been discussing are slight. They usually consist of a single amino acid substitution. For example, it is possible to account for a large part of the naturally occurring variability among the chains of the human hemoglobin molecule by assuming that changes in the amino acid sequence reflect a change in a single DNA base pair in the structural gene in question. Thus, as compared to the corresponding "normal" chains, each of 34 different alpha chains, 56 beta chains, 4 gamma chains and 4 delta chains, all of which are known to exist, can be accounted for by single base pair changes.

Mutations that are produced by more extensive changes in DNA molecules are also known to occur. For example, altered DNA molecules can result from unequal crossing over (crossing over usually occurs shortly before cells undergo meiosis, a form of cell division discussed in Chapter 1). The result of unequal crossing over is that hybrid protein molecules can be produced. Thus persons whose hemoglobin molecules contain a protein chain with an amino acid sequence that begins like that of a normal delta chain and ends like that of a normal beta chain have been reported. These unusual molecules probably result because of unequal crossing over between the structural genes for the human beta and delta chains. (Very recently, it has been suggested that unequal crossing over may also have played a part in generating the repeated segments of DNA that have been reported within the genetic material of most higher organisms.)

Mutations can also result from physical damage to DNA molecules. Localized regions of DNA molecules can be damaged by energy in the form of radiation of various kinds, including ultraviolet light, X-rays, cosmic rays, and various types of natural radiation from radioactive substances. In order to compensate for the damaging effects of certain kinds of radiation, organisms have evolved enzymes that repair radiation-damaged segments of DNA. For example, as shown in Figure 4-12, the main way in which ultraviolet light damages DNA is by causing adjacent thymine bases to become tightly bonded to one another, thus distorting the molecule and interfering with its ability to replicate itself. Human cells usually contain an enzyme that has the rather remarkable ability to excise the abnormal regions of DNA, after which other enzymes then repair the excised segment. Rare persons are deficient in the enzyme that does the excising of the abnormal radiation-damage segments. As expected, their skin is highly sensitive to ultraviolet light and may be severely damaged on exposure to sunlight. (This rare trait is autosomal recessive.)

DNA molecules can also be damaged by heat and by various kinds of chemicals. But before we discuss the relative importance of these and other factors in the production of human mutations, we must first mention what happens to mutated genes in the human population once they have arisen.

A person whose DNA contains a mutated gene may be affected positively, negatively, or to no detectable degree. If the influence is negative, the affected person is usually less likely to reproduce than an unaffected person, or may even die before reaching reproductive age. In either instance the mutation will be lost from the population unless it recurs spontaneously. If the mutation has no detectable effect on the person, then its persistence in the population is a matter of chance. As we have seen, this is probably true of the great majority of polymorphisms involving protein molecules, such as that for the protein chains of the hemoglobin molecule discussed in preceding paragraphs. Finally, if the mutation has a positive effect it will tend to become more prevalent in a population with the passage of time, mostly because of natural selection, though chance can still play a role. The allele for sickle-cell hemoglobin and its relation to malaria provide a good example.

But almost all mutations are either frankly detrimental or have no detectable effect. This is not surprising. Organisms are already so precisely adapted to survival and reproduction in the environments in which they live that any change is much more likely to be a disadvantage than an advantage. In protein polymorphisms, a high degree of diversity is probably maintained in part by chance and in part by natural selection, though the relative roles of each of these factors in most structural differences of human proteins have yet to be explained.

Estimating the Mutation Rates of Human Genes

Mutations are rare events, and it is particularly difficult to estimate the frequency with which they occur in human populations. There are two main ways, known as the *direct* and *indirect* methods, of estimating the rate at which human genes mutate. Both methods are subject to many sources of error and can yield at best, approximate estimates. (Both methods apply only to traits that are obvious and detrimental.)

The direct method of determining the rate at which human genes mutate requires finding out about all occurrences of a particular dominant disorder and determining how many of them appear sporadically among the offspring of unaffected parents. As you may recall, about 15 percent of the cases of Marfan's syndrome occur in this way (see Chapter 1), and data from an obstetrical hospital in Copenhagen indicate that mutation may account for a much larger

4–13
Brother and sister achondroplastic dwarfs of the Owitch family as they appeared in 1949 when they arrived in Israel after having spent several years in Auschwitz concentration camp. Their lives were spared because they were used for medical experiments. The autosomal dominant gene (or genes) responsible for this trait does not impair fertility. (Courtesy United Press International.)

percentage of instances of achondroplastic dwarfism (Figure 4-13).

The more obvious sources of error in the direct method are these. First, it is hard to be sure that *all* of the instances of any disorder have been identified, no matter how apparent the disorder may be. Second, it is possible that the trait being studied may result from any of several different mutations, all of which produce the same end result. (This is probably true for achondroplastic dwarfism.) Third, a dominant gene may be incompletely expressed in a particular person because of its interaction with other genes. Thus a particular trait may suddenly appear in the offspring of parents who, although they appeared unaffected, actually carried the gene for the disorder. Of course, it would then not be a mutation.

Although the direct method applies only to dominant traits, the indirect method of estimating the mutation rate of human genes can be applied to both dominant and recessive traits. The indirect method is based on the assumption that the rate at which mutant genes are added to a population is balanced by the rate at which they are removed by natural selection. This implies that affected persons are less likely to reproduce than those who are unaffected, which tends to

TABLE 4–2
Estimates of mutation rates of certain human genes from normal to abnormal.

TRAIT	MUTANT GENE PER 100,000 SEX CELLS
Autosomal dominants	
Huntington's chorea	<0.1
Nail-patella syndrome	0.2
Epiloia (type of brain tumor)	0.4–0.8
Aniridia (absence of iris)	0.5
Retinoblastoma (tumor of retina)	0.6–1.8
Multiple polyposis of the large intestine	1–3
Achondroplasia (dwarfness)	4–12
Neurofibromatosis (tumors of nervous tissue)	13–25
X-Linked recessives	
Hemophilia A	2–4
Hemophilia B	0.5–1
Duchenne-type muscular dystrophy	4–10

reduce the frequency of the mutant gene and to numerically cancel out the number of new mutant genes added by spontaneous mutation during any time interval. The indirect method depends on estimating *reproductive fitness,* which is a measure of the likelihood that a person affected by a particular trait will reproduce. The indirect method is subject to the same sources of error as the direct method, and the estimation of reproductive fitness thus adds still another source of error, which makes the indirect method of estimating human mutation rates even less reliable than the direct method.

When all of these factors are taken into account, the most reliable estimate is that a mutation for any particular human gene, or at least for those genes that result in rare, detrimental traits in their mutated forms, is encountered in about one out of every 100,000 sex cells. For the most part, mutations seem to occur about equally often among the sex cells of both sexes. Table 4-2 shows estimates of the mutation rates of certain human genes.

From the total number of autosomal recessive traits that are known and from the rate at which at least some human genes responsible for autosomal recessive traits are known to mutate, it can be calculated that all people probably carry at least several highly detrimental autosomal recessive genes. This estimate is borne out by the rate at which autosomal recessive traits appear among the offspring of consanguineous matings. Overall, it is estimated that the average person probably carries about five recessive genes that in the homozygous condition could result in death before the person could reproduce.

Although the mutation rates for individual genes are somewhat variable, higher mutation rates for almost all genes can be produced in several ways. For example, increased amounts of radiation, higher temperatures, and the effects of various chemicals all tend to increase the mutation rate, sometimes markedly so. (These factors directly alter DNA molecules, but their effects oftentimes show up as damage to entire chromosomes, or to parts of chromosomes. Phenotypic changes that occur because of altered chromosomes are also considered mutations. See Chapter 1.)

It has been estimated that technologically advanced populations, because of exposure to diagnostic X-rays, radioactive fallout, and natural sources of radiation, have tripled the amount of radiation to which their sex cells are exposed during the reproductive years. But there is evidence that overall, radiation from X-rays, fallout, and other sources does not play a major role in the production of human mutations, at least those with apparent effects.

On the other hand, it is known that for various experimental animals the spontaneous mutation rate increases directly with temperature, as is expected of any process that can be explained in molecular terms. This may have some relevance for human genetics because it has been found that the temperature of the gonads of men wearing pants, especially tight-fitting ones, is higher than that of men wearing kilts or wearing nothing at all.

Various chemicals are known to increase the mutation rate among experimental animals, including fruit flies and mice. Nitrogen mustard is a good example, and because of its biochemical effects on rapidly dividing cells, this toxic substance has been of some benefit in the treatment of certain human cancers. Many other chemicals that many of us encounter in our everyday lives—caffeine, for example—are known to increase the mutation rate in bacteria and in some insects, but evidence that they do so in human beings is not conclusive. Nonetheless, in the end most human mutations probably result from short-lived, highly reactive chemicals that are produced inside normal living cells. These highly reactive chemicals are usually formed as the byproducts of normal metabolism, or, more rarely, may result from the effects of radiation.

In recent years it has been asked whether manufactured chemicals, which are so readily available in the environments of industrialized populations, are increasing the rate at which our species' genes are mutating. While environmental chemicals may have had some influence on the mutation rate of some human genes, it is impossible to estimate their overall effect at this time. Mutation rates comparable to those estimated for human beings have also been observed among fruit flies, mice, and other creatures raised under controlled laboratory conditions. It may be true that mutation rates in the human population are not influenced to any presently measurable degree by manufactured chemicals, but rather depend in large part on spontaneous chemical reactions that occur inside normal cells. Nonetheless, it seems wise to continue to question whether or not the benefit of introducing a manufactured chemical into a particular environment warrants the risk that the substance will damage the genetic material of the human and nonhuman populations living there.

Summary

The protein molecules of individuals of the same species are extremely variable. Slightly different forms of almost all human proteins are known to exist, and most of the time slight differences in protein molecules from one person to another do not have ill effects. For most human protein polymorphisms there is no evidence that natural selection influences the frequencies of the slightly different genes that code for the slightly different proteins. Nonetheless, the gene for the abnormal beta chain of hemoglobin that is characteristic of sickle-cell anemia is maintained at a rather high frequency because natural selection favors the reproduction of heterozygotes in malarious areas.

The worldwide distribution of the ABO blood groups is probably best explained as the result of human migrations. Although various blood groups have been correlated with ulcers, with resistance to certain infectious diseases, and with maternal-fetal incompatibilities, the evidence that natural selection plays a major role in how the ABO blood groups are distributed is not convincing.

In large populations random fluctuations in the frequencies of those genes not obviously influenced by natural selection tend to cancel each other. But in small populations, significant changes in genetic constitution often occur by chance alone. Thus the Dunkers, who have been reproductively isolated for many generations, have gene frequencies that are different from those of either their West German forebears or their American neighbors.

Most widespread species look slightly different in different geographic areas, and this reflects some degree of genetic difference between populations. Distinct local breeding groups are known as races. In practice most of these local populations blend with one another because many individuals in neighboring populations interbreed.

External features of the human body, such as skin color and body build, are generally more variable from one local human population to another than are less apparent traits, such as slight differences in protein molecules. Racial differences in body surfaces probably became established in ancient, isolated human groups in large part because of natural selection. The number of human races recognized depends on the purposes and opinions of the person who classifies them.

Although human races surely exist, they are for the most part biologically irrelevant today. This is because most people no longer adapt to the environment primarily by means of body surfaces, but rather by means of language and culture. Human racial differences have been blurred in recent years because of increased mobility and social changes.

Mutations are heritable changes that arise within existing DNA molecules, most of them because of some kind of accident. Mutations that arise during DNA replication usually are produced by the substitution of a single base pair. Unequal crossing-over, radiation of various kinds, heat, and numerous chemicals can either produce mutations directly or speed up the rate at which they occur.

Mutations in the human population are particularly hard to detect. Direct and indirect methods of estimating the mutation rate of certain human genes have been devised, but they are subject to many sources of error, and the estimates are very approximate.

Exposure to X-rays, environmental chemicals, natural radiation, and heat surely have some effect on the rate at which human genes mutate, but they are probably not the most important factors. Rather, intrinsic errors in the replication of DNA as well as the presence of short-lived by-products of normal metabolism produced inside normal cells probably account for the occurrence of most human mutations.

Suggested Readings

1. "The Genetics of Human Populations," by L. L. Cavalli-Sforza. *Scientific American,* Sept. 1974. Discusses how the molecular differences within human populations are greater than those between populations.
2. "Sickle Cells and Evolution," by Anthony C. Allison. *Scientific American,* Aug. 1956, Offprint 1065. How natural selection can maintain an allele that seems to be frankly detrimental.
3. "The Genetics of the Dunkers," by H. Bently Glass. *Scientific American,* Aug. 1953, Offprint 1062. Describes the evidence for genetic drift among the members of a religious sect.
4. "Genetic Drift in an Italian Population," by L. L. Cavalli-Sforza. *Scientific American,* Aug. 1969. Discusses the genetic effects of physical isolation and of consanguineous marriages among people who live in Italy's Parma Valley.
5. "Ionizing Radiation and Evolution," by James F. Crow. *Scientific American,* Sept. 1959, Offprint 55. Discusses the role of X-rays and other kinds of ionizing radiation in the evolutionary process.

This painting by Pablo Picasso ("Girl before a Mirror," 1932, March 14. Oil on canvas, $64 \times 51\frac{1}{4}''$) symbolizes the fragmented, emotional consideration that the human species sometimes gives to its own genetic future. (Collection, The Museum of Modern Art, New York. Gift of Mrs. Simon Guggenheim.)

CHAPTER 5

Genes and Human Intervention

Although our understanding of genetics began with the publication of Mendel's manuscript in 1866, people have known how to use it to their advantage since ancient times. Consider the case of "man's best friend," the domestic dog *(Canis familiaris).* Archaeological evidence suggests that the ancestors of domestic dogs were frequent visitors to the camps of prehistoric peoples 30,000 years ago and that the dog had been fully domesticated by 12,000 years ago: its remains are regularly found close to those of human beings from that time on. Our prehistoric forebears must have noticed that some of their canine companions had more desirable characteristics than others and that parents sometimes passed on desirable (and undesirable) physical and behavioral traits to their offspring. People took advantage of this observation by selective breeding—that is, by the deliberate mating, generation after generation of those animals that had the most desirable traits. By means of selective breeding it was possible to produce different *breeds* of dogs that not only looked different from one another but also varied in behavioral traits, such as the degree of tameness, the tendency to bite or to bark, and the willingness to obey commands.

At the present time there are about 110 officially recognized breeds of dogs, almost all of which were developed by means of controlled matings (Figure 5-1). All of these breeds are members of the same species because all dogs can interbreed, or at least can exchange genes through intermediates. (Physical

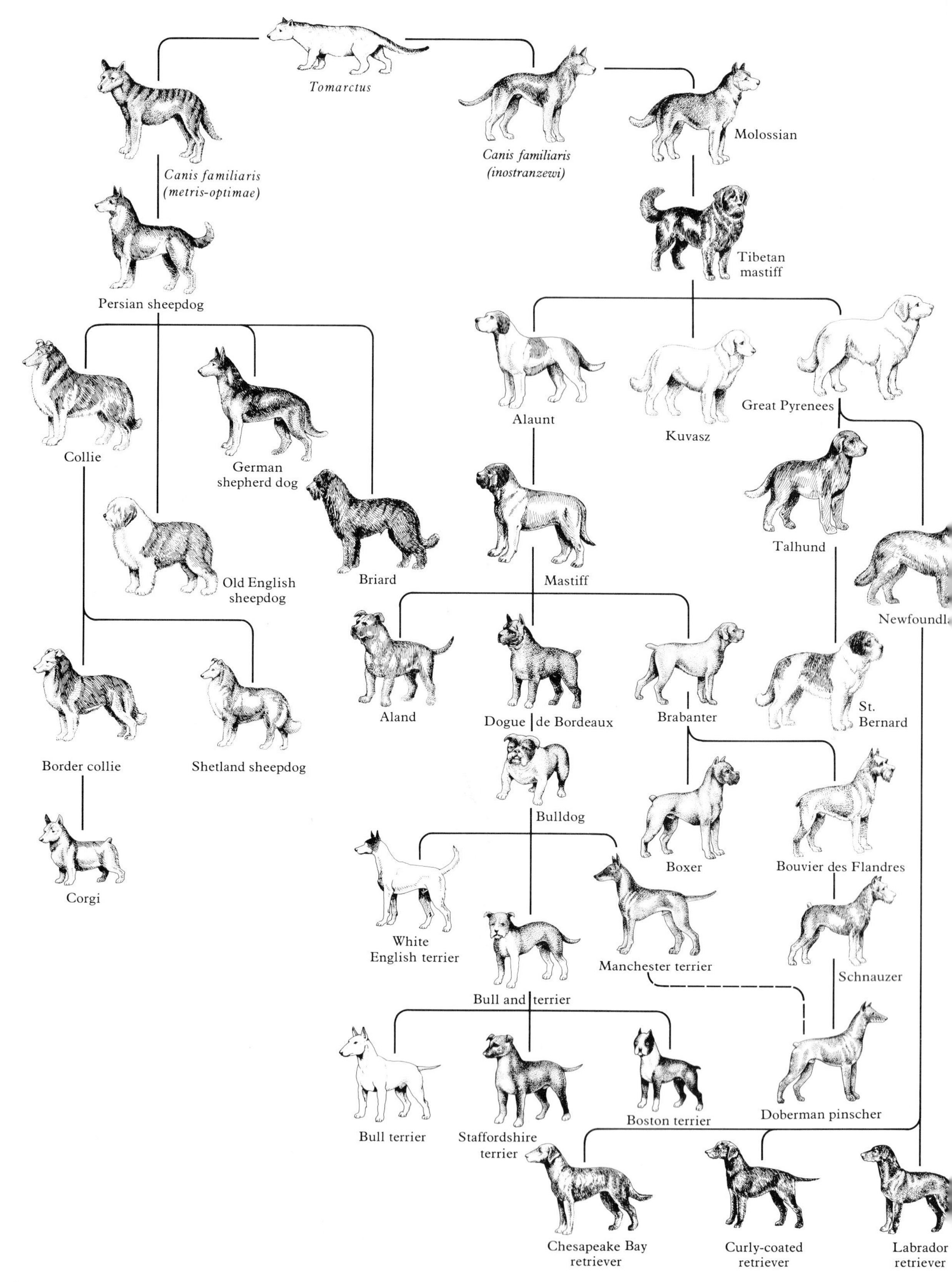

5–1
Some of the major lineages of modern dog.

differences obviously prevent Saint Bernards from directly mating with Chihuahuas, but the two can exchange genes indirectly by mating with breeds of intermediate size.) Dogs are much more variable than most other species of mammals, and this variability is clearly a product of human intervention. The fact that selective breeding has produced the remarkable diversity of domestic dogs (and of other domestic plants and animals) has raised an important thought in human minds from at least Plato's time, and probably from long before. About 2,500 years ago Plato suggested that it might be possible to improve the human species by controlling human variability by means of selective breeding. Plato was probably not the first person to make this suggestion, and he was certainly not the last.

The term *eugenics* was first applied to programs or proposals for the improvement of the human species by means of selective breeding, but the meaning of the term has broadened with technological advances in recent years. To simplify a little, *negative* eugenics has as its goal decreasing the propagation of "undesirable" traits within the human population, and *positive* eugenics is concerned with increasing the propagation of "desirable" traits, not only by selective breeding, but by other means as well.

The obvious problem of eugenics in any form is that in practice it depends on the definition of "desirable" and "undesirable" that is applied to human traits. Sometimes traits are obviously undesirable in that they cause those who have them to be at a severe disadvantage, perhaps to die before reaching reproductive age. But for many variable human characteristics, including most protein polymorphisms, there are no obvious ill or advantageous effects and the terms "desirable" and "undesirable" are meaningless. And, as we shall see, it is often far from clear what is desirable or undesirable in the genetic aspects of human behavior.

A further difficulty of eugenics in any form is that the genetic component of many human traits is complex and incompletely understood. This is especially true of some of the most interesting human traits, including certain features of behavior that are known to have a complicated genetic basis. As we shall discuss in following paragraphs, it is impossible to assess the effects of the environment on the expression of many traits, especially behavioral ones, that are known to be genetically influenced.

In this chapter our main concern is with proposals and programs that have been suggested for genetically improving the human species. We shall first discuss the effects of prenatal diagnosis and genetic counseling and then consider the prospects for, and some of the possible genetic consequences of, what has become known as *genetic engineering.* Then we shall turn our attention to the genetic basis of certain aspects of human behavior. Specifically, we shall discuss the genetics of IQ scores in human populations and the behavior of males who are of sex chromosome constitution XYY. You will see how the environment can strongly influence human behavior. Finally, we shall discuss ways in which natural selection is at work in human populations today and make some cautious speculations about how our species may evolve in the future.

But a word of caution is in order. These are delicate and heady subjects. At best, the application of eugenic measures to human populations could result in the elimination of some frankly detrimental traits or in the widespread distribution of advantageous ones. But at worst, the application of supposed eugenic measures could result in the irrational horrors of genocidal wars or in other kinds of deplorable human activities. The subject of eugenics cannot be discussed with complete objectivity—the issues are too important and too emotion-charged. In particular, anyone who writes about the topic bears the burdens (and the blessings) of a certain background, genes, and political opinions, and this is bound to color any presentation, however careful, of the facts about eugenics. The practice of eugenics arouses opinions, emotions, and sometimes political action. The controversies surrounding the subject will undoubtedly continue for a long time to come.

Let us begin our discussion of eugenics by turning our attention to some traits that show simple Mendelian patterns of inheritance and that could therefore readily lend themselves to eugenic measures.

Genetic Counseling and Prenatal Diagnosis

In earlier chapters we discussed the patterns of inheritance of some human diseases and abnormalities whose genetic basis lies in a single pair of alleles. You will recall that such disorders may be transmitted in either a dominant or a recessive fashion and that their alleles may be located on autosomes or on sex chromosomes. Genetic abnormalities of this sort have simple Mendelian patterns, so reliable predictions can therefore be made about the likelihood that two parents of known genotype will produce an affected child. It is then up to the parents to decide whether the risk is substantial enough to influence their decision about whether to reproduce (see Table 5-1).

In itself genetic counseling is not a form of eugenics, but it becomes one if people who have, or are carriers of, a particular Mendelian trait choose not to reproduce because of the genetic counseling they receive. If they do, genetic

TABLE 5–1

"Recurrence risks" are based on statistical data and on genetic theory. They have been calculated for over 500 known or suspected genetic conditions. The examples show the risk of the birth of an additional affected child to parents who already have one affected offering.

MAGNITUDE OF RISK (%)		GENETIC BASIS
Total	100	Both parents are homozygous recessives
High	75	Both parents are heterozygous for an autosomal dominant with full penetrance
	50	One parent is heterozygous for an autosomal dominant
	50	For sons, a sex-linked gene carried by the mother
Moderate	30	Down's syndrome due to translocation of part of a 21 to another autosome in one parent
	25	Recessives with full penetrance; both parents heterozygous
Low	5 or less	Down's syndrome due to trisomy 21, arising from meiotic nondisjunction in one parent, most likely the mother

Source: After A. G. Motulsky and F. Hecht.

counseling is a form of negative eugenics, because it has the effect of reducing the frequency of a particular abnormal allele in the human population. But although genetic counseling can and does help to alleviate human misery by sparing some parents the financial and emotional expenses of caring for an affected child, genetic counseling has had little effect on our species' overall genetic make-up, at least so far.

To see why this is so, consider the example of *galactosemia,* an autosomal recessive defect of carbohydrate metabolism whose most serious clinical manifestation, if untreated, is severe mental retardation. This serious inborn error of metabolism resembles phenylketonuria (PKU) in two ways: it, too, is caused by an enzyme defect and the mental retardation can be prevented by dietary restriction, in this case, of milk sugar. Galactosemia occurs in about 25 out of every million children born in the United States, and it is estimated that about 10,000 people per million (1 in 100) in the U.S. are actually heterozygous carriers of the disease. (The observed frequency of the disease is less than the high carrier rate might lead us to expect. In large part this is probably because many fetuses affected by galactosemia die before birth.) Biochemical screening tests to detect heterozygotes are available, so it would be possible to identify all carriers of galactosemia in the United States by a mass screening program.

The fact that heterozygotes can be identified makes it theoretically possible to eliminate the abnormal allele for galactosemia from the U.S. population by means of genetic counseling alone. But consider what this would entail. First, assuming a U.S. population of 200 million, there are 2 million carriers of galactosemia in this country, and all those of reproductive age would have to be identified by means of costly screening tests. Second, all carriers of reproductive age and younger would have to agree never to reproduce simply because they were carriers of galactosemia. (When considering the likelihood that a heterozygous carrier would choose not to reproduce, bear in mind that if you live in the United States your chances of being a carrier are one in 100.)

Most people would agree that the benefits of preventing the birth of 25 galactosemic children per million does not warrant either the expense of nationwide screening or the sacrifice of the reproductive potentials of the 10,000 people per million who happen to be carriers. Besides, you will recall that based on the mutation rates of various human alleles and on the frequency with which severe genetic abnormalities are observed among the offspring of consanguineous matings, it is estimated that each of us probably carries several abnormal recessive alleles that would be lethal in the homozygous state. In the extreme, if we could devise screening tests for and then detect all carriers of serious autosomal recessive disorders, and if all of the carriers chose not to reproduce because they were carriers, the end result would be that nobody would reproduce, and our species would become extinct.

The main effect of genetic counseling so far, and probably in the future, too, is that some parents (and other people in affected family lines) are spared the financial and emotional strains of having a child who is severely affected by a predictably inherited trait. But at this time genetic counseling has little overall effect on the genetic constitution of the human species, and it is not likely to have an effect for a long time to come. (We will discuss some of the undesirable effects of the medical treatment of genetic diseases later in this chapter.)

In recent years the effectiveness of genetic counseling has been greatly increased because of the availability of the medical procedure known as *amniocentesis.* This diagnostic technique consists of collecting a sample, by means of a needle carefully inserted through the abdomen of a pregnant woman, of the amniotic fluid that surrounds the developing fetus (Figure 5-2). Cells from the sample are then grown in tissue culture and subsequently analyzed for biochemical (or chromosomal) defects. Also, the fluid itself may be subjected to various biochemical tests. Table 5-2 is a list of some of the inborn errors of metabolism that can be detected by amniocentesis. At the present time the procedure is usually performed during the middle three months of pregnancy; the technique is generally safe and is becoming increasingly accurate. Amniocentesis is very effective in detecting chromosomal abnormalities such as Down's syndrome. In fact, the detection of abnormal chromosome constitutions is at present the most common reason for which amniocentesis is performed. As discussed in Chapter 1, greater maternal age is strongly associated with the increased incidence of Down's syndrome. Pregnant women who are 40 years old or older are usually advised to have the procedure performed, provided that they are willing to abort the development of a fetus that has an abnormal number of chromosomes.

TABLE 5–2
Some of the inborn errors of metabolism that can be detected by amniocentesis. Note that most of these errors are associated with mental retardation.

DISORDER	DEFFECTIVE ENZYME OR METABOLIC DERANGEMENT
ASSOCIATED WITH MENTAL RETARDATION	
Chromosomal abnormalities (Down's syndrome, Turner's syndrome, XYY, etc.)	Excess or deficiency of total genetic information
Arginosuccinic aciduria	Arginosuccinase
Citrullinemia	Arginosuccinate synthetase
Fucosidosis	Alpha-fucosidase
Galactosemia	Galactose-1-phosphate uridyl transferase
Gaucher's disease	
Infantile type	Absent cerebrosidase
Adult type	Deficient cerebrosidase
Generalized gangliosidosis	Absent beta galactosidase
Juvenile GM_1 gangliosidosis	Deficient beta galactosidase
Juvenile GM_2 gangliosidosis	Deficiency of Hexosaminidase A
Glycogen storage disease type 2	Alpha-1, 4-glucosidase
Hunter's disease	Increased amniotic fluid Heparitin sulfate
Hurler's disease	Increased amniotic fluid Heparitin sulfate

TABLE 5–2 (continued)
Some of the inborn errors of metabolism that can be detected by amniocentesis. Note that most of these errors are associated with mental retardation.

DISORDER	DEFFECTIVE ENZYME OR METABOLIC DERANGEMENT
ASSOCIATED WITH MENTAL RETARDATION	
I-cell disease	Multiple lysomal hydrolases
Isovaleric acidemia	Isovaleryl CoA dehydrogenase
Lesch-Nyhan syndrome	Hypoxanthine-guanine-phosphoribose transferase
Maple syrup urine disease	Alpha-keto isocaproate decarboxylase
Metachromatic leucodystrophy	
Late infantile type	Absent arylsulfatase A
Juvenile and adult types	Deficient arylsulfatase A
Methylmalonic acidemia	Methylmalonyl CoA carbonyl mutase
Niemann-Pick disease	Sphingomyelinase
Refsum's disease	Phytanic acid alpha-oxidase
Sandhoff's disease	Hexosaminidase A and B
Sanfilippo disease	Increased amniotic fluid Heparitin sulfate
Tay-Sachs disease	Hexosaminidase A
Wolman's disease	Acid lipase
POSSIBLY ASSOCIATED WITH MENTAL RETARDATION	
Cystathioninuria	Cystathionase
Homocystinuria	Cystathionine synthase
NOT ASSOCIATED WITH MENTAL RETARDATION	
Adrenogenital syndrome	Increased amniotic fluid corticosteroids
Cystinosis	Increased cellular cystine
Fabry's disease	Alpha-galactosidase
Hypervalinemia	Valine transaminase
Orotic aciduria	Orotidylic pyrophosphorylase and orotidylic decarboxylase

Source: From "Prenatal Diagnosis of Genetic Disease," by T. Friedmann.

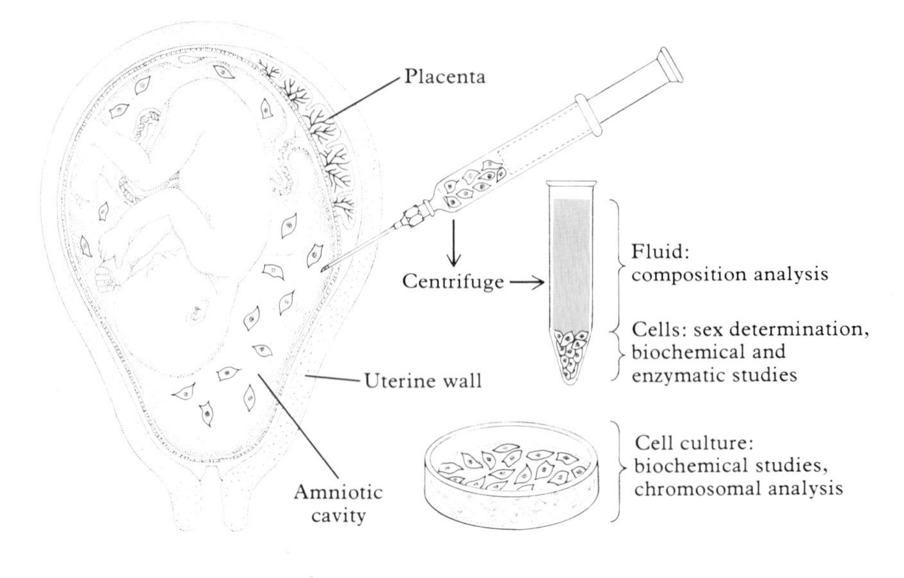

5–2
Cells and amniotic fluid can be obtained for biochemical and chromosomal analysis by means of amniocentesis. (From "Prenatal Diagnosis of Genetic Disease," by Theodore Friedmann. Copyright © 1971 by Scientific American, Inc. All rights reserved.)

Like all other human undertakings that can be considered eugenic in the broad sense of the term, amniocentesis is vigorously and emphatically opposed by some people. At the present time, if a fetus is proved to be affected by a trait detectable by means of amniocentesis, the only recourse other than allowing the birth of a severely affected child is abortion. The practice of aborting fetuses for any reason is repugnant and utterly unacceptable to some people, who therefore oppose amniocentesis because they consider it the first step toward abortion, which it sometimes is. It is hoped that abortion will not always be the only alternative. It will probably be possible someday to treat some genetically determined abnormalities of developing fetuses (such as inborn errors of metabolism) early in pregnancy by correcting the defective DNA that ultimately gives rise to it. Although we are at present a very long way from being able to directly manipulate the genetic material of developing human fetuses, the mid-1970s saw some spectacular (and to some people disturbing) advances in our ability to directly alter the genetic programs of various living things. The direct human manipulation of an organism's genetic program usually is called genetic engineering. It is no longer a futuristic notion, but rather a controversial reality that is worth discussing further.

Genetic Engineering

There are three main ways in which an organism's genetic program could be manipulated: by changing the DNA already present or by either adding to or subtracting from it. At the present time almost all genetic engineering is accomplished by adding new DNA fragments to an existing genetic program, and so far most of the research has been concerned with the introduction of foreign genes into bacterial cells. One technique for doing this depends on the existence of structures called *plasmids,* which are short double-stranded circular pieces of DNA that are found inside certain bacterial cells and that replicate independently of the bacterial chromosome (Figure 5-3). By making use of appropriate enzymes, it is possible to generate fragments of purified DNA molecules from almost any species and then to splice some of the fragments into

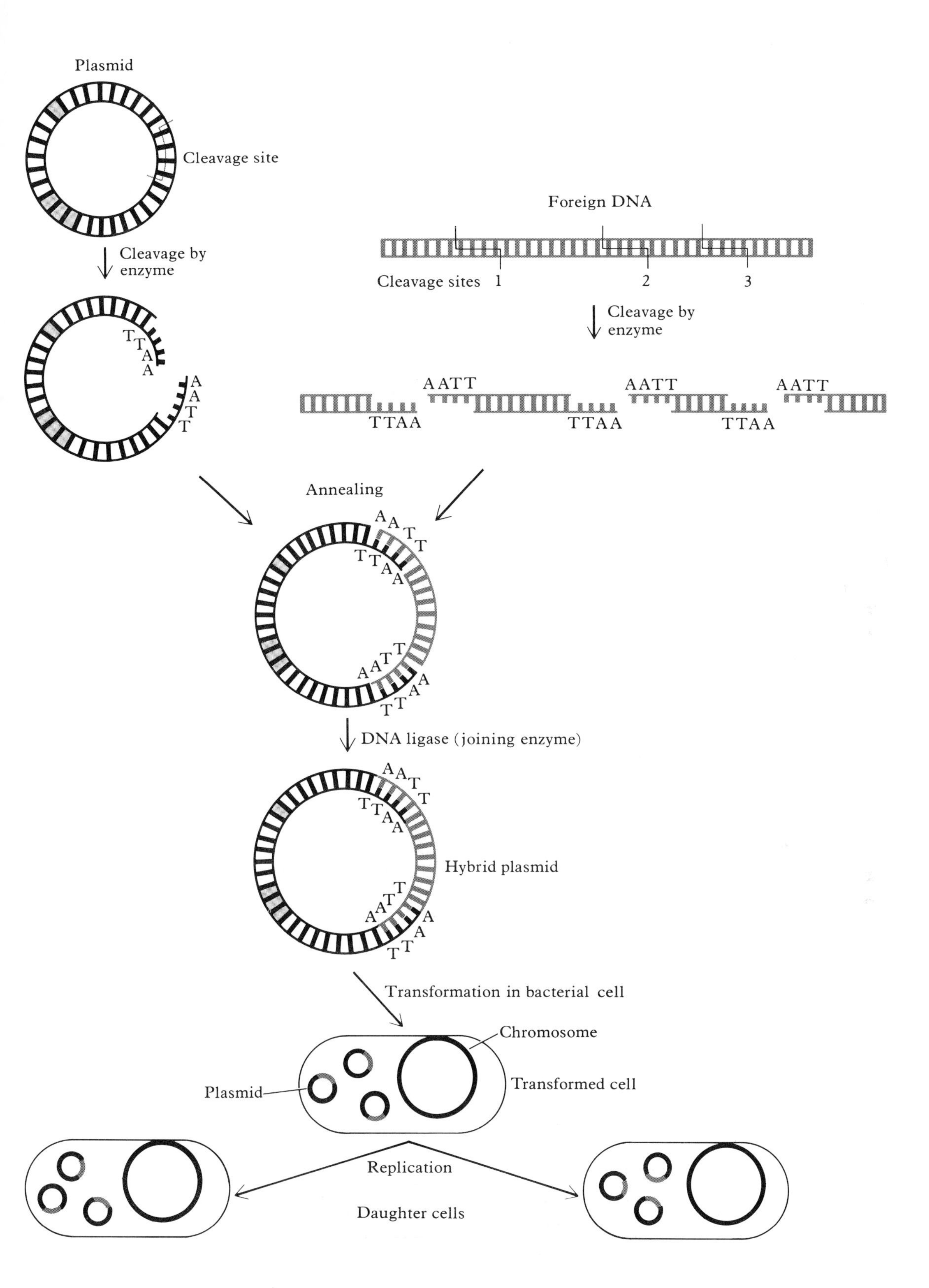

5–3
Foreign DNA can be spliced into a plasmid and introduced with the plasmid into a bacterium. The foreign DNA is replicated along with the bacterial chromosome shortly before cell division occurs. (From "Manipulation of Genes," by Stanley N. Cohen. Copyright © 1975 by Scientific American, Inc. All rights reserved.)

a plasmid that can then be put back inside a living bacterium. (The most commonly used bacterium is *Escherichia coli (E. coli),* a widespread organism that is regularly found in great abundance in the human large intestine.) When a hybrid plasmid replicates, the fragments of foreign DNA may be replicated too, and if these fragments happen to contain a gene not normally found in *E. coli,* it is sometimes possible to detect the corresponding gene product. In other words, it is sometimes possible to splice a foreign gene into a bacterium and to recover from the bacterium the biochemical product of that gene.

At present, the state of the art of genetic engineering is limited almost exclusively to introducing foreign genes into bacterial cells. But there is reason to expect that in the not-too-distant future techniques for incorporating certain genes into human and other mammalian cells will be developed. In fact, this may have already been accomplished, not once, but twice—one time spontaneously and the other intentionally. The spontaneous incident concerned a form of the enzyme arginase that is not usually present in human cells. The blood of scientists who had worked with a certain virus for many years was recently reported to contain some unusual arginase activity. Presumably, this is because some of the scientists' body cells became infected with the virus, whose genetic program is known to contain the gene for the form of arginase in question. (It is known that rabbits intentionally infected by the same virus acquire the capacity to synthesize this enzyme.) It has also been recently reported that cells from a person who has galactosemia can acquire the capacity to synthesize the enzyme they usually lack (galactose-1-phosphate uridyl transferase, or G1PUT) if the cells are infected by a virus carrying a gene that codes for normal G1PUT. The implications of these reports are clear: it may be possible before too long to treat some of those human diseases that result from defective enzymes (or other proteins) by directly introducing into an affected person's cells a particular virus whose DNA normally contains, or has had spliced into it, the normal version of the gene that is defective.

If and when geneticists overcome the technological difficulties of regularly bringing about the manufacture of particular gene products inside the cells of an affected person, what will the impact on human genetics be? To simplify a little, there would probably be little or no positive impact on future generations. This is because at the present time genetic engineers are directing their attention to correcting genetic defects in body cells (somatic cells), not sex cells. To continue with our example, suppose it becomes possible to regularly bring about the manufacture of G1PUT inside the cells of people who have galactosemia. That is surely to be desired, but it is of no consequence to the genetics of our species unless the gene for G1PUT is somehow incorporated into the genetic program of *sex* cells of those who are affected. Technologically, it would be extremely difficult to first detect and then correct, by genetic engineering or any other means, human sex cells that are defective for a particular allele. Therefore, the undeniable good effects of relieving a person of the symptoms of galactosemia by means of genetic engineering would have no positive or negative eugenic effects, because the sex cells would still contain the defective allele.

On the other hand, genetic engineering could have an *adverse* effect on human genetics in that affected individuals would be made healthier and perhaps would be more inclined to pass on their abnormal alleles to future generations. Of course, the modern medical treatment of genetic diseases could have the same detrimental effect. But the accumulation of unfavorable alleles in a large and widespread population is a very slow process, and most geneticists agree that, although human activities may already have increased the frequency of some abnormal alleles in some segments of the human population, any effects on the worldwide population have probably been negligible, at least so far. Moreover, it is possible that the potential adverse eugenic effects of genetic engineering and of modern medical treatment could be offset by the potential favorable effects of genetic counseling. This would occur if the affected people, whose lot is made easier by some kind of human intervention, felt obligated not to pass on their abnormal alleles and therefore chose not to reproduce, or to adopt children. But there is no guarantee that all affected people would elect this course of action, and in the end we may be doing our species a genetic injustice by devising oftentimes elaborate treatments for genetically determined disorders. Whether or not these concerns are grounds for not rendering effective treatment, either through medical practices or through genetic engineering, is left for you to consider.

Although its overall eugenic effects would probably not be of much consequence, the social and ethical problems that could arise from the practice of genetic engineering are formidable and unprecedented. Many geneticists recently showed their concern over a related issue by imposing upon themselves a moratorium on some aspects of research into genetic engineering that could potentially have devastating, though largely unpredictable, effects on people and on other living things. Their concern was that the manipulation of certain genes would give rise to new kinds of bacteria whose properties might allow them to infect people and other creatures with serious, perhaps fatal, effects. The moratorium has now been lifted, but only after serious, sometimes heated, debate in international meetings convened by geneticists in order to collectively discuss how to proceed with their provocative—yet potentially dangerous—research. The result of these international meetings was a set of guidelines for carrying out potentially dangerous experiments in genetic engineering that could result in the release of novel infectious organisms. Also important, the guidelines were devised, not only by geneticists and other scientists, but also by persons trained in sociology and ethics. It is to be hoped that scientists will continue to acknowledge publicly that they lack expertise in nonscientific matters and that guidelines concerning research on genetic matters that have possible eugenic overtones will result from decisions arrived at not only by professional scientists and professional humanists, but by all of us who are concerned with the genetic future of the human species.

The scientific, ethical and political aspects of traits that show simple Mendelian patterns of inheritance are sometimes in conflict, and traits whose genetic basis is more complex are the subject of even greater controversy. The problem is that some of our species' most familiar and engaging traits do not depend on a single pair of alleles, but rather on many that interact with one another and with the environment in ways that are complicated and poorly understood. Included among such traits are certain aspects of human behavior. But before we discuss the genetic component of human behavior and its important relationship to environment, we must first consider some of the technical difficulties of trying to sort out the genetic and environmental factors that cooperate to make each one of us unique. In other words, we must now discuss the effects of nature (genetic factors) and nurture (environmental factors) as they relate to human uniqueness.

Nature and Nurture—The Concept of Heritability

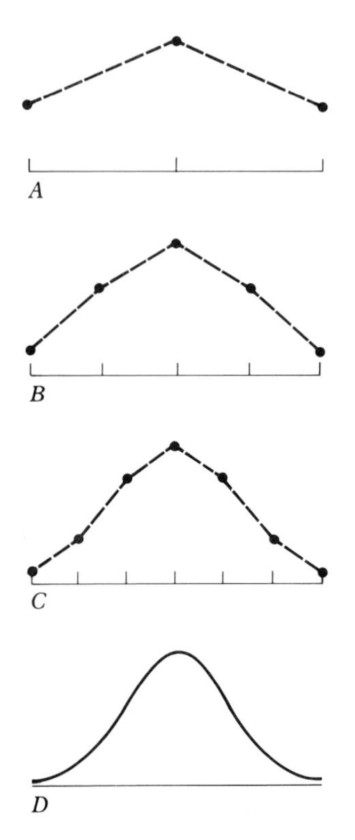

5–4
How the frequency distributions of phenotypes are related to the number of pairs of alleles involved. A, one pair of alleles distributed over three phenotypes. B, two pairs of alleles distributed over five phenotypes. C, three pairs of alleles distributed over seven phenotypes. D, an infinite number of pairs of alleles distributed over a continuous array of phenotypes. (This curve is the normal distribution curve *that characterizes most biological populations. Compare with Figure 1-14.) (From Curt Stern,* Principles of Human Genetics, 3rd Ed. *W. H. Freeman and Company. Copyright © 1973.)*

As discussed in preceding chapters, the inheritance of those traits that show continuous and gradual variation within a population usually depends, not on a single pair of alleles, but on many pairs that in each person interact both with the rest of the individual's genes and with the environment. Consider, for example, the multitude of genetic and environmental factors that interact to determine how tall a person is. Height is influenced by many genes, such as those that code for growth hormone, for intestinal enzymes that digest food and thus provide the body with building blocks for growth, and for the rate at which calcium is deposited in the long bones of the legs. But environmental factors also make an important contribution to a person's stature. For example, the absence of sunlight can result in inadequate synthesis of Vitamin D, which may result in the slowing down of proper bone growth. Chronic poor nutrition in childhood can also influence adult height. The end result of all of these factors affecting height, and of the many others that must also play a part, is that in the worldwide human population, and in any random sample of it we wish to single out, height is "normally" distributed. That is, most individuals in any human population are not far away from being of some "average" height, and although there are very tall and very short people, there are fewer of them than people of average stature (Figure 5-4).

Plant and animal breeders have known for centuries that continuously varying traits are influenced by the environment to different degrees. Because the breeders are concerned with establishing true-breeding lines of plants and animals that have desirable characteristics, it is important for them to be able to assess the relative effects of genes and environment on characteristics they consider desirable, such as copious milk production from cows, large eggs from hens, and long coats on woolly sheep. All of these characteristics are influenced by many genes and by many environmental factors, and breeders have found the statistical concept of *heritability* of use in assessing the relative role of the genetic factors that they would like to "breed into," and thereby genetically improve, the breeds that they have already developed. Heritability is that proportion of the phenotypic variation in any population that can be attributed to genetic factors.

How does one go about assessing the relative effects of genes and environment on a continuously varying trait that is desirable? In practice it is very difficult to estimate heritability, but there are several ways of going about it. For example, heritability can be estimated by mating individuals from a given position in the normal distribution curve for a particular trait, and then examining the distribution of the trait in their offspring, who must be raised under strictly controlled environmental conditions. If all of the variation in a particular trait in a particular population is due to genetic factors alone, the heritability is assigned the number 1 (or 100 percent), and the average value of the trait in the offspring is equal to the value of the position on the curve from which the parents were selected. On the other hand, if all of the variation is due to environmental factors, the heritability is 0 (0 percent), and the average value of the trait in the offspring is the same as the average value in the population from which the parents were selected. Most continuously varying traits have heritabilities between 0 and 1 that can be estimated by determining the difference between the average values in selected parents and their offspring (Figure 5-5).

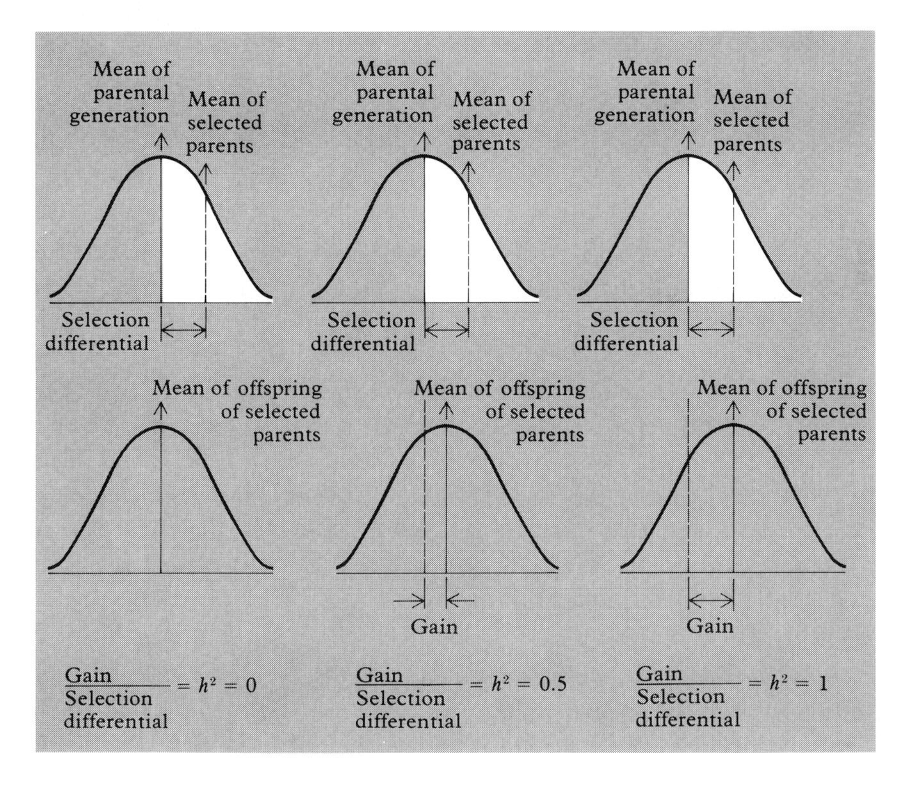

5–5
Heritability can be estimated by breeding experiments that allow one to calculate the ratio of the "gain" to the "selection differential." In the three top curves the difference between the mean (average) of the selected parents and the mean of the population they were selected from (the selection differential*) is the same. In each case the offspring that are produced have characteristics that form a normal distribution curve, and the change in mean between parents and offspring (the* gain*) can be used to estimate the heritability of a particular trait, which is represented by* h^2*. (From I. Michael Lerner and William J. Libby,* Heredity, Evolution, and Society, *2nd Ed. W. H. Freeman and Company. Copyright © 1976.)*

Heritability can also be estimated by experiments in inbreeding, over at least several generations and under controlled environmental conditions, of animals that are related. This method of estimating heritability depends on the fact that different kinds of relatives (brothers and sisters, cousins, etc.) share to different degrees the genes that influence the trait under consideration. Figure 5-6 shows the range of heritabilities, as determined largely by experiments in inbreeding, for various economically important traits of the domestic chicken.

But heritability is a statistical concept that applies to populations, not to individuals. For example, as shown in Figure 5-6, the average heritability for the weight of hens' eggs is about 0.75 (75 percent). A heritability of 0.75 does not mean that for each hen egg weight is determined three-fourths by heredity and one-fourth by environment. What it does mean is that overall three-fourths of the total variation in the weight of hens' eggs is associated with genetic differences, and the remaining one-fourth of the total variation is associated with differences in environmental factors.

Although heritability is never easy to measure, it is particularly difficult to estimate in human populations. There are several reasons for this, the most obvious of which is that people are not experimental animals that can be selectively mated, highly inbred, or raised under strictly controlled conditions. A less obvious reason why heritability is difficult (in fact, impossible) to determine accurately in human populations is that it depends, not only on genes and environment, but on the *interaction* between the two. In human populations, information concerning the exact contribution that the interaction of genes and environment make to the heritability of human traits is in many instances meager, if not nonexistent.

Nonetheless, the heritability of certain human abnormalities that depend on many genes can be crudely estimated from the rate at which the abnormality

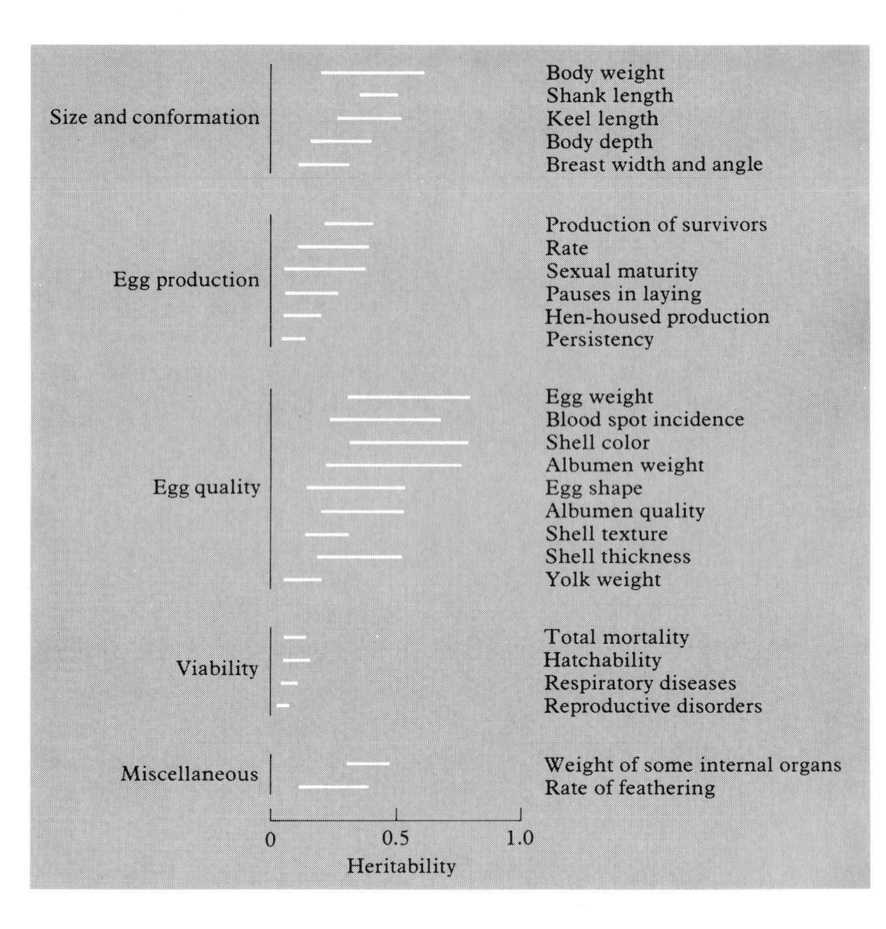

5–6
The range of heritabilities reported for various economically important characteristics of the domestic chicken. (From I. Michael Lerner and William J. Libby, Heredity, Evolution, and Society, *2nd Ed. W. H. Freeman and Company. Copyright ©1976.)*

occurs among close relatives of an affected person, compared with the population at large. Thus it is roughly estimated that the heritability of hydrocephalus (water on the brain) is about 0.4 (40 percent), whereas that for certain kinds of epilepsy is about 0.5 (50 percent). Similarly, the heritability of clubfoot is estimated to be about 0.8 (80 percent), and that of hairlip (with or without cleft palate) is roughly 0.7 (70 percent).

Some biologists and statisticians have recently suggested that the concept of heritability, because of its built-in limitations, should not be applied to human populations at all. There is probably some virtue to this argument, especially for estimates of poorly defined or poorly understood traits such as "intelligence" (see the following discussion). But fortunately, nature has provided us with at least one fairly accurate way of estimating the effects of genes and environment in human populations, whether one has much faith in the concept of heritability or not. Human beings sometimes are produced in almost duplicate copies known as identical twins, who for all practical purposes are genetically identical.

Nature and Nurture—Twin Studies

There are two kinds of human twins: one-egg, or *monozygotic,* twins and two-egg, or *dizygotic,* twins. Dizygotic twins are produced when two eggs, rather than the usual one, are released at ovulation and both are subsequently

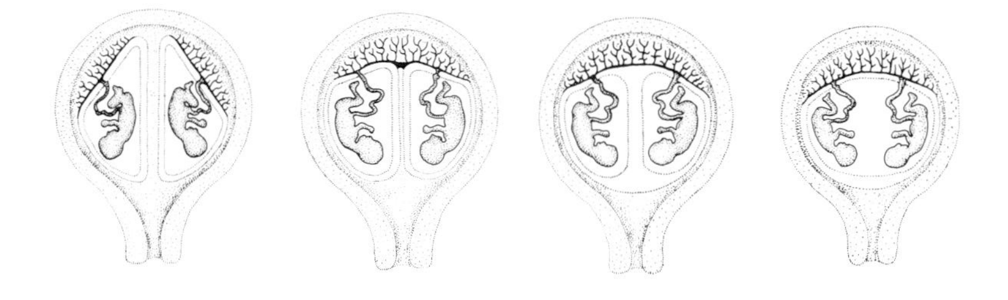

5–7
Monozygotic and dizygotic twins may share the prenatal environment to different degrees because of the arrangement of the membranes surrounding the fetuses.

fertilized by different sperm. The genetic relationship between dizygotic twins is thus the same as between brothers and sisters who are not twins. On the other hand, monozygotic twins orginate from a single fertilized egg that, after having been fertilized by a single sperm, splits into two at a very early stage of development and thus results in two individuals who are, barring mutations during development and in later life, genetically identical (Figure 5-7).

Identical (monozygotic) twins have all of their genes in common, so any variation between two identical twins is in large part due to the effects of the environment. Most of the time identical twins are raised in the same environment, and under such circumstances they tend to strongly resemble each other both physically and in at least some aspects of behavior (Figure 5-8). But sometimes monozygotic twins are raised apart from one another in different environments, and when this happens a study of the differences between the two individuals can provide a fairly accurate measure of the genetic and environmental components of certain human characteristics.

Statistical studies of the differences between identical twins sometimes focus on complex characteristics that depend on many genes and many environmental factors but that nonetheless do not show continuous and gradual variation. The relative role of genetic factors in the expression of such all-or-none traits can be estimated by the degree to which the twins are concordant or discordant. When both twins have a particular trait they are concordant, and when only one does, they are discordant. Table 5-3 shows the degree to which monozygotic and dizygotic twins are concordant for various abnormal conditions that depend on the interaction of many genes and many environmental factors. The percentage of concordance provides an estimate of the degree to which a particular condition is genetically influenced.

Of special interest in studies of the differences between monozygotic twins reared apart are differences in behavior and the degree to which these differences are directly influenced by genes. But before we turn our attention to the genetics of human behavior, we should first discuss some of the difficulties of trying to figure out the genetic basis of some of those traits that impart to our species its complexity and its uniqueness.

The Genetics of Human Behavior—Some Difficulties

The most variable thing about the human species is the endless variety of ways in which people behave. Our species is unique among all living things in that its members adapt to the environment primarily by means of a complicated form of learned behavior called culture, which is transmitted from generation to generation by means of the symbol system of language. As you may recall, the biological basis of these complicated learned behaviors lies in the human brain,

5–8
Top, identical twins Bruno and Giorgio Schreiber at about two years of age. Even the adult twins cannot tell who is who in this photo. Center, as adults, the twins continued to look very much alike. Bottom, a recent photo. Bruno (left) and Giorgio (right) are now professors of zoology at the Universities of Parma (Italy) and Belo Horizonte (Brazil), respectively. (Courtesy of B. Schreiber.)

which is the most complicated organ that evolution has produced so far. The capacities for language and culture are shared by all normal human beings, and insofar as they depend on the organization of the human brain the capacities for language and culture may be considered genetically determined.

But very little is known of how the behavior of individuals relates to the brain. How the contributions of genetic and environmental factors interact throughout the course of a person's life to result in behavior is largely unknown to us. Nonetheless, recent years have brought some advances in our understanding of the genetic basis of certain abnormalities in behavior, and these are

TABLE 5–3
The percentage of concordance among twins for some traits that depend on many genes and many environmental factors.

OBSERVED DISEASE OR BEHAVIOR	PERCENTAGE CONCORDANCE	
	MZ TWINS	DZ TWINS
Tuberculosis	54	16
Cancer at the same site	7	3
Clubfoot	32	3
Measles	95	87
Scarlet fever	64	47
Rickets	88	22
Arterial hypertension	25	7
Manic-depressive syndrome	67	5
Death from infection	8	9
Rheumatoid arthritis	34	7
Schizophrenia (1930s)	68	11
Criminality (1930s)	72	34
Feeble-mindedness (1930s)	94	50

Source: From *Heredity Evolution and Society,* 2nd ed., by I. Michael Lerner and William J. Libby. W. H. Freeman and Co. Copyright © 1976.

worth discussing further. Let us now consider some behavioral abnormalities known to result from the presence of a single defective gene.

You will recall that both PKU and galactosemia result from homozygosity for autosomal alleles that code for defective enzymes and that both of these conditions produce mental retardation if they are untreated. (Many genetic defects are known to result in some degree of mental retardation, and this bears witness to the intricacy of the human brain's architecture and biochemistry. Nonetheless, mental retardation can also result from wholly environmental factors, such as accidents during birth and chronically inadequate nutrition of the fetus. See Figure 5-9.) In both PKU and galactosemia the abnormal enzyme presumably leads to a biochemical defect in the developing brains of affected persons, and this defect results in mental retardation, though exactly how this occurs is poorly understood. PKU and galactosemia would thus seem to be examples of a behavioral abnormality that is wholly genetic and independent of environmental factors. Yet environmental factors, in the form of special diets, can prevent the mental retardation associated with both of these conditions. Thus, environmental factors sometimes strongly influence a person's behavior even when genetic factors appear to be of overwhelming importance.

An intriguing human ailment that depends on the presence of a single defective allele and that results in abnormal behavior is the *Lesch-Nyhan syndrome,* first described in 1965. The abnormal allele for this fatal condition is located on the X chromosome (sex-linked) and its presence results in a deficiency for the enzyme hypoxanthine-guanine phosphoribosyl transferase (HGPRT). Only males are affected; affected females die as embryos. A rather shocking fact about this fortunately rare disease is that affected boys have abnormal posture and spastic, involuntary movements (among other abnor-

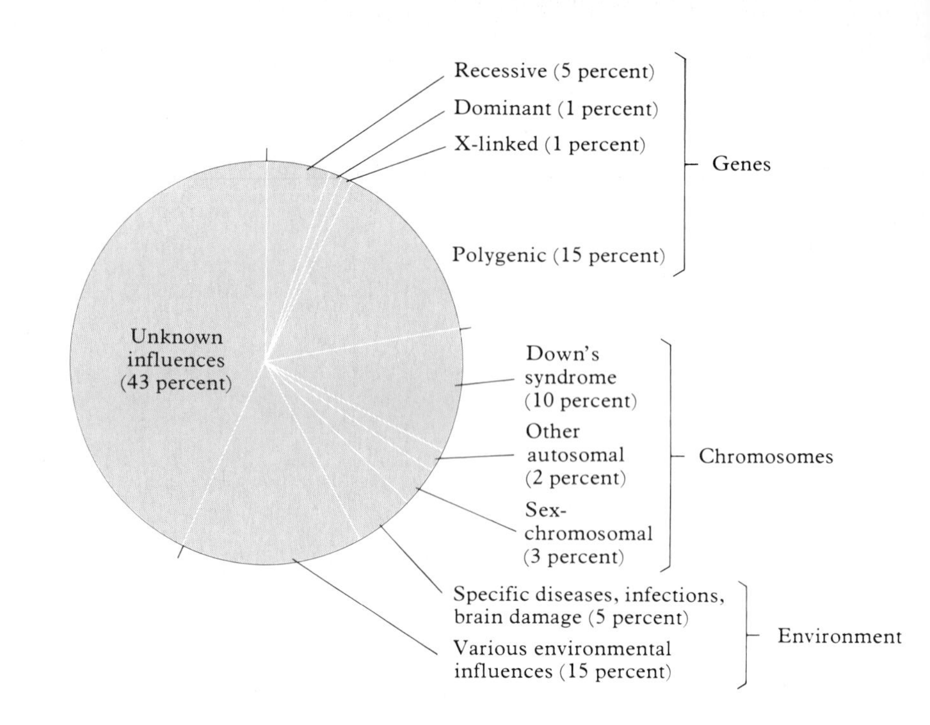

5–9
Estimates of the relative incidence of types of mental retardation that result from genetic and environmental factors. (After Penrose, in Wendt (ed.), Genetik und Gesellschaft, *Wiss. Verlagsges., Stuttgart, 1971.)*

malities), and they regularly engage in a bizarre form of self-destructive behavior. Unless their teeth are removed, they invariably bite off the tissue of their lips, and they frequently use their teeth to tear at the flesh of their hands, sometimes mutilating themselves severely. (The self-mutilation is thought to be involuntary, because they appear to be terrified by their self-destructive activities.) Although the exact relationship between the absence of the enzyme HGPRT in certain parts of the brain and the presence of this unfortunate behavior remain to be explained, the Lesch-Nyhan syndrome is an example of abnormal behavior that as far as we know is determined by simple genetic factors alone.

Genetic factors are also of obvious importance in producing the abnormal behaviors manifested by many (but not all) people who have abnormal numbers of chromosomes. The mental retardation associated with Down's syndrome (trisomy-21) is one example, and, as shown in Figure 5-9, this relatively common disorder accounts for about one-tenth of the total of the people in the United States (about 3 percent of the population) who have some form of mental retardation. On the other hand, the behavior of many people who have abnormal sex chromosome constitutions, such as XO, XXX, and XXY, is apparently normal in spite of their chromosomal abnormalities. (We will discuss the possible behavioral effects of the chromosome constitution XYY later in this chapter.) Overall, the study of the behavioral effects of abnormal numbers of human chromosomes suggests that normal mental activities depend on the effects of many genes, located on many chromosomes, and this is borne out by the rate at which severe mental retardation occurs among the offspring of matings between close relatives.

A good example of an abnormality of human behavior that probably depends on the interaction of many genes and a multitude of environmental factors is *schizophrenia.* Contrary to popular usage of the word, this serious behavioral abnormality, of which several types are officially recognized, is not manifested by "split personality," but rather by a split between thoughts and feelings, and by a loss of contact with the environment (there are many other features). Schizophrenia is by no means rare. It is estimated that in the 1970s at least 2 million people in the United States alone either have schizophrenia or will develop symptoms of the disease sometime during their lives.

Although the issue is far from settled, most geneticists are convinced that genetic factors influence the development of certain types of schizophrenia. Part of the evidence for this comes from studies of monozygotic twins raised apart. As shown in Table 5-4 both develop schizophrenia much more commonly than do both dizygotic twins who are raised in different environments.

But what about the genetics of human behavioral traits that are neither so obvious nor so extreme as severe mental retardation, self-destructive activities, or schizophrenia, but that could nonetheless be of some importance to the future evolution of the human species? For example, what is the genetic basis of the tendency for compassion, for altruism, for perseverence, for leadership, for realism, for wisdom, for curiosity, or for any of the seemingly endless qualities of human behavior that can be singled out, labeled, and measured by some kind of psychological test? As of the late 1970s it has yet to be determined whether most measurable behavioral differences between human beings result primarily from genetic or environmental factors. Nonetheless, there is every reason to expect that such complex and sometimes ill-defined behavioral traits result from the interaction of many genes and a great variety of environmental factors.

A further difficulty in assessing the genetic basis of many aspects of human behavior is that a person's behavior rarely remains the same for any extended period of time. Most of us do not behave the same way we did 10 years ago (or 20 or 30 years ago), and few of us will be behaving as we do now 10 or 20 years hence. Overall, the genetic basis of any human behavior is not only complex, hard to define, and difficult to measure, but changeable as well.

In the end, we have little evidence about the genetic basis of the behavior of individual people, and it is most often impossible to accurately assess the influence that the environment has on it. Nonetheless, the relative contributions of genes and environment to human behavior are sometimes of considerable social interest, in large part because of eugenic proposals that bear on human behavior. We now turn our attention to a human characteristic that is socially important in spite of the fact that its genetic basis is poorly understood and that it is subject to a great number of environmental influences: intelligence.

TABLE 5–4
Concordance of schizophrenia in twins.

COUNTRY	YEAR	MONOZYGOTIC TWINS		DIZYGOTIC TWINS	
		NUMBER OF PAIRS STUDIED	PERCENT CONCORDANCE	NUMBER OF PAIRS STUDIED	PERCENT CONCORDANCE
Denmark	1965	7	29	31	6
Germany	1928	19	58	13	0
Great Britain	1953	26	65	35	11
Japan	1961	55	60	11	18
Norway	1964	8	25	12	17
United States	1946	174	69	296	11

Source: From *Heredity, Evolution and Society,* 2d ed., by I. Michael Lerner and William J. Libby. W. H. Freeman and Co. Copyright © 1976.

The Genetics of IQ Scores

Intelligence quotient, or IQ, is best defined, not as a measure of a person's "intelligence," but as a measure of a person's ability to perform well on IQ tests. This definition is appropriate because it is impossible to define intelligence to everyone's satisfaction, and even if we could define the term, we could not measure it by the same yardstick in all human populations. (To get some idea of the difficulties, how would you measure the intelligence of European college students as compared with that of people of the same age group who live in the Amazonian rain forest, where they must survive without the benefits of the technological achievements of the culture of the former group?) IQ tests were developed to measure some of the mental aptitudes of white middle-class people who live in the United States and in Europe. IQ test results are informative in that they provide a reliable way of predicting success in those environments, as judged by standards considered appropriate to those particular cultures. Thus the degree to which a person is able to score high on IQ tests is important only insofar as it is important to achieve success as measured by academic, occupational, social, and other standards within a given cultural framework. Judgements of what constitutes success are always open to question and subject to change. Whatever its ultimate importance, within white middle-class populations the ability to score high on IQ tests appears to be strongly influenced by genetic factors.

As usual, the main problem in trying to figure out the genetic basis of IQ scores is that it is impossible to accurately sort out the genetic and environmental influences that interact to determine how a person performs on IQ tests. Nonetheless, there are several ways of roughly estimating the genetic component of IQ scores. Monozygotic twins that are raised apart are one source of information, and studies of them indicate a high degree of heritability. (For traits that show continuous and gradual variation, whether between twins or among other relatives, the degree of similarity is best measured by the value of the statistical quantity known as the *correlation coefficient,* whose numerical value can vary between 0 and 1. In general, a high value of the correlation coefficient suggests a high heritability. See Figure 5-10.)

5–10
A summary of correlation coefficients concerning IQ scores compiled by L. Erlenmeyer-Kimling and L. F. Jarvik from various sources. The horizontal lines show the range of correlation coefficients for "intelligence" between persons who are related to various degrees either by genes or by environment. (From I. Michael Lerner and William J. Libby, Heredity, Evolution, and Society, *2nd Ed. W. H. Freeman and Company. Copyright © 1976.)*

Genetic and nongenetic relationships studied			Genetic correlation	Range of correlations (0.00 0.10 0.20 0.30 0.40 0.50 0.60 0.70 0.80 0.90)	Studies included
Unrelated persons	Reared apart		0.00		4
	Reared together		0.00		5
Foster-parent-child			0.00		3
Parent-child			0.50		12
Siblings	Reared apart		0.50		2
	Reared together		0.50		35
Twins	Two-egg	Opposite sex	0.50		9
		Like sex	0.50		11
	One-egg	Reared apart	1.00		4
		Reared together	1.00		14

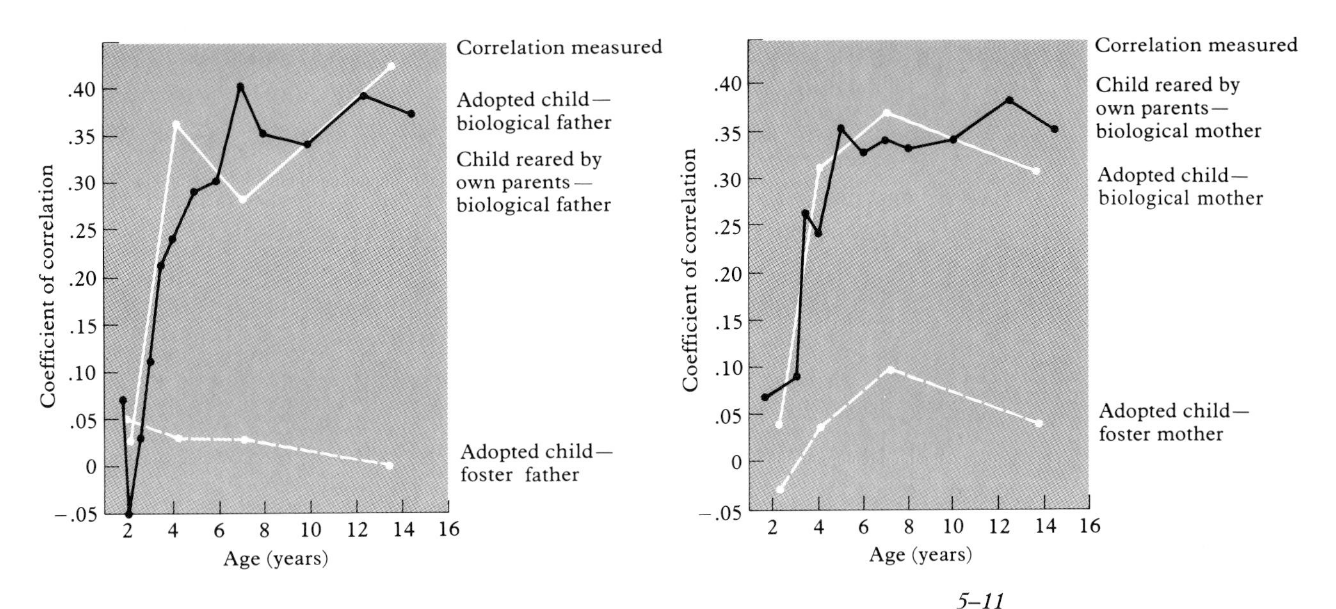

5–11
Correlations between the IQ of children and the education of biological and foster parents. (From M. Honzik, and M. Skodak and B. M. Skeels.)

The genetic component of IQ scores can also be estimated by studies of adopted children. As shown in Figure 5-11, the IQs of parents and their offspring are directly related, whether the offspring are raised by their biological parents or not. On the other hand, the IQs of adopted children are not affected by those of their foster parents.

The heritability of IQ scores as estimated in different populations and by various methods ranges from about 0.3 (30 percent) to 0.9 (90 percent) and this wide range of estimates serves as a reminder that heritability is impossible to accurately measure in human populations. Nonetheless, most studies give values of about 0.5 or 0.6 (50 to 60 percent): this implies that about 40 or 50 percent of the total variation in IQ scores within the population being studied can be attributed to the interaction of genes and environment, or to environmental factors alone.

We have already discussed the role of nutrition in the development of normal mental capacities, and many other environmental factors are known to affect, not only normal mental development, but IQ scores as well. Among the most important environmental components are psychological and social factors, and several are known to influence IQ scores. First, as shown in Figure 5-12, there is evidence that offspring of larger families tend, overall, to have lower IQ scores than those from smaller families. This has been interpreted (though not without criticism) as evidence that large families may provide an environment that is less satisfactory to the development of higher IQ scores. Second, some incomplete but ongoing studies of the effects of the environment within a given social class indicate that the presence of an "enriched," as opposed to a "deprived," environment may favor the development of higher IQ scores. This refers to the fact that most children who have lower IQ scores come from homes in which the kinds of mental activities measured by IQ scores are not emphasized. And third, the children of people who have higher occupational status tend to have higher IQs than children whose parents are of lower occupational status. As shown in Table 5-5, this is true of the populations studied in the United States, Russia, and England.

Although the reasons for the variation observed between the IQ scores of whites of different socioeconomic classes are poorly understood and sometimes hotly disputed, the controversy surrounding them is dwarfed by the ignorance

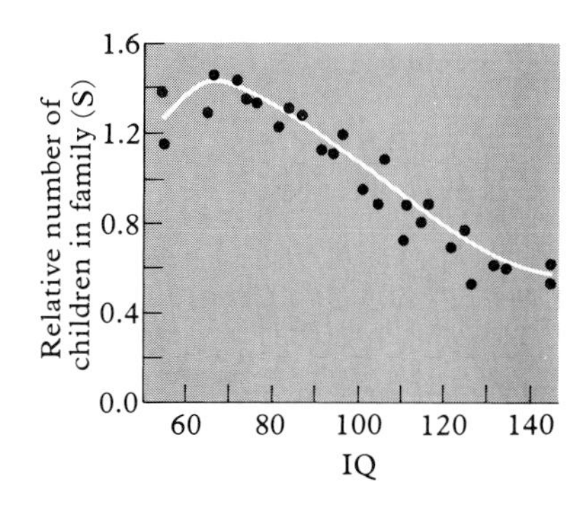

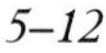

5–12
Data concerning the relation of family size and average IQ. The relative number of children in each family is the actual number of children divided by the average number of children per family in the whole sample. (From Human Diversity, *by Kenneth Mather, Free Press, 1964.)*

TABLE 5–5

Data about the relation between occupational status of parents and the average IQ of biological and adopted children.

OCCUPATIONAL STATUS OF PARENTS	AVERAGE IQ OF CHILDREN					AVERAGE SIZE OF ENGLISH FAMILY
	U.S.	USSR	CHICAGO		ENGLAND	
			ADOPTED	BIOLOGICAL		
Professional	116	117	113	119	115	1.73
Semiprofessional	112	109	112	118	113	1.60
Clerical and	107	105			106	1.54
retail business			111	107		
Skilled	105	101			102	1.85
Semiskilled	98	97	109	101	97	2.03
Unskilled	96	92	108	102	95	2.12

Source: From *Heredity Evolution and Society,* 2d ed., by I. Michael Lerner and William J. Libby. W. H. Freeman and Co. Copyright © 1976.

and confusion surrounding the average IQ scores of various human races, particularly those of blacks and whites in the United States. As shown in Figure 5-13, the difference between the average IQ scores of blacks and whites in the U.S. population is about 15 IQ points. (The graph shows the distribution of the IQ test scores of white school children of all social classes as sampled in 1960 from across the United States, compared with the scores of black school children from Alabama, Florida, Georgia, Tennessee, and South Carolina in 1963.) A glance at the figure reveals that there is a difference between the average IQ scores of blacks and whites raised in the United States. The question is: what does this difference mean, and how important is it?

Some white academicians, among whom Arthur Jensen and William Shockley are perhaps the most vocal, have asserted that most, if not all, of the differences in IQ scores between black children and white children can be attributed to genetic factors. But it is in fact impossible to sort out the exact contributions of genetic and environmental factors to IQ scores *within* either the black or the white American populations, so it is clearly impossible to make meaningful comparisons *between* the two. No matter how high the heritability of IQ scores actually is, the fact remains that, given the obvious cultural differences between populations of middle-class whites and ghetto-dwelling blacks, we would *expect* to find a difference in their performances on IQ tests. In the face of vast environmental differences, it seems premature at best to attribute the expected IQ score difference entirely to systematic genetic differences between the two populations.

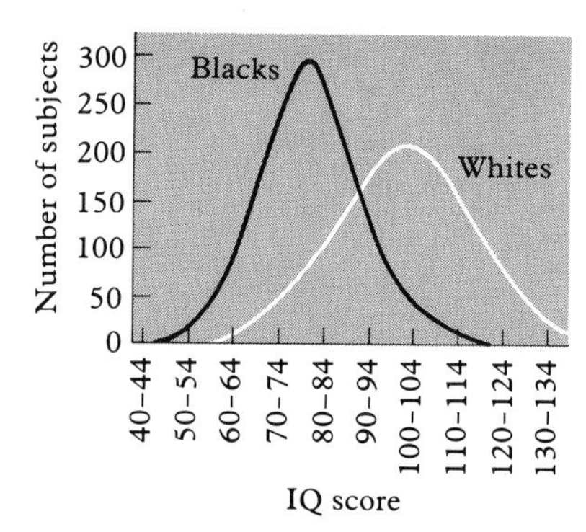

5–13
The distribution of IQ test scores of white school children from all social classes as sampled in 1960 across the United States, as compared with the scores of black school children from the schools of Alabama, Georgia, Florida, Tennessee, and South Carolina in 1963. (From "Intelligence and Race," by W. F. Bodmer and L. L. Cavalli-Sforza. Copyright © 1970 by Scientific American, Inc. All rights reserved.)

In the opinion of many geneticists, the black–white differences in IQ scores most likely reflect, not genetic, but rather cultural and other environmental factors, such as nutrition, family size, and the psychological impact of racism. In this regard it should be noted that differences in IQ between whites of upper and lower socioeconomic groups are somewhat larger than the IQ differences between American blacks and whites. Also, environmental factors probably vary more between than within these two groups, and IQ tests are culturally biased toward middle-class whites. But this is not to say that genetic factors may not influence IQ differences between races. The point is that most of the differences in IQ scores between American blacks and whites could probably be eliminated by changing the social environments, including the schools, so that they become the same for both groups. Finally, IQ tests only predict a person's success in a middle-class academically oriented environment. They are limited, as is any human endeavor, by the knowledge and experiences of those persons who devise them. In the end, IQ scores measure an aspect of human behavior that, like most other features of the ways in which people behave, is desirable in the opinion of some people and undesirable in the opinion of others.

We now turn our attention to an aspect of human behavior that is clearly undesirable to most people, and that has been said by some to be associated with the presence of an extra Y chromosome. We will now discuss the relationship between criminal behavior and the chromosome constitution XYY.

The Behavior of Men Whose Sex-Chromosomes are XYY

We have already discussed the way in which males whose sex chromosomes are XYY may be produced from various accidents that occur during cell division. XYY males are born with surprising frequency. It is estimated that about one in every 1000 newborn males has XYY sex chromosomes. Most XYY men are taller than those whose chromosomes are XY and some of them have severe acne, but most of them appear otherwise normal.

The controversy that has surrounded the XYY genotype concerns criminal behavior. In 1965 it was reported that the genotype XYY was encountered among men in a certain wing of a "mental-penal institution" (namely, the Carstairs maximum security hospital in Scotland) at a much higher rate than among the general population. (Seven out of 197 men were XYYs, which is about 36 times their frequency in the population at large.) Since that time, other well-documented studies have been published from around the world. The conclusion, though hotly contested by some, seems inescapable to others. It is this: Men of sex-chromosome constitution XYY are somewhat more likely to be incarcerated in a "mental-penal institution" than men whose sex chromosomes are XY. (In general, the offenses of XYY men are similar to those of XYs, and contrary to some earlier reports, XYY men do not appear to be concentrated among the most dangerous, aggressive, or violent inmates.)

Nonetheless, the great majority (at least 98 percent) of men in mental-penal institutions have XY sex chromosomes, and only a small proportion of the total number of men who have sex-chromosome constitution XYY engage in criminal behavior. Based on the percent of XYY men in mental or penal institutions, compared with the incidence of XYY males in the general population, it is estimated that at least 96 percent do not behave in ways that result in their being institutionalized.

Criminal behavior is so widespread and it affects the lives of so many noncriminals that the relation between a slightly increased tendency for criminal behavior and the presence of an extra Y chromosome may be of more than academic interest. But at the same time the issue is so surrounded by prejudice and concerns such important social considerations that there is good reason to argue that it ought not be investigated further. As usual, the problem is that it is impossible to accurately assess the effects of the genetic versus the environmental factors. Environmental factors, such as growing up in a ghetto or associating with people who commit criminal acts, are of great importance in the development of criminal behavior.

Now consider some of the moral and ethical difficulties that would arise in the course of a genetic screening program designed to detect XYY males in the general population. Suppose you have just identified an XYY male infant. What would you advise his parents? What would you expect the parents to do because of your advice? Would you tell the child that he is an XYY and thereby implant in his mind the notion that he may grow up to be criminal? Would you mind if the chromosomes of your own male offspring were routinely screened for the presence of an extra Y chromosome without your knowing about it?

As of May 1976 you need not give much urgent thought to these matters, at least if you live in the United States. The only XYY screening program in the United States has, at least for the present, been shut down because of unrelenting pressure from people opposed to XYY screening. In the spring of 1975 the faculty of Harvard Medical School, brought to caucus by geneticists and psychiatrists interested in continuing a screening program that had been in existence since 1968, voted 200 to 30 to continue the screening project. Nonetheless, a few months later the screening project was shut down because those in charge of the project said they were worn out by the pressures of some activist groups that opposed XYY screening.

How should we proceed in this delicate area that may or may not have some genetic basis and that may or may not be subject to potential eugenic measures? Clearly, that depends on the judgements of individuals, and opinions concerning the matter are abundant and oftentimes strongly felt. Perhaps it would be appropriate to follow the example of molecular geneticists who—when recently confronted by the potentially disastrous effects of their research involving hybrid DNA molecules—convened a series of international conferences to discuss the problem. Any such conference on XYY chromosomes should probably include not only geneticists, sociologists, criminologists, social philosophers, and other humanists, but also people of sex chromosome constitution XYY and representatives of well-informed groups of people that have strong interest in or opposition to XYY screening.

As if all of the uncertainty surrounding the genetics of IQ scores and the behavior of XYY males were not enough, we now turn our attention to a brief consideration of how natural selection may affect human populations in the future, as it does today. (In the discussion of future human evolution that follows, you should remember that, although speculation is sometimes fascinating, sometimes dangerous, and sometimes rewarding, in the end it is just speculation.)

How Natural Selection at Work Today May Affect Human Evolution in the Future

Like all other living things, human beings have evolved and are evolving. But understanding how evolution works does not allow us to make predictions about the future course of the evolution of our own species, or of any other. Nonetheless, we can do some cautious, perhaps meaningful, speculating about the future evolution of the human species, provided that we base our speculations on the assumption that natural selection is at work in the human population today and will be in the future.

How Natural Selection at Work Today May Affect Human Evolution in the Future

You will recall that natural selection is at the core of the theory of evolution, and that in essence it consists of differential reproduction within populations in which individuals differ from one another genetically. Because of natural selection, individuals best suited to survival in a given environment leave more descendants than those who are not so well suited, and because of the effects of natural selection, living things are precisely adapted to the environment. It has been argued that natural selection is no longer important in the evolution of the human species because few human beings now live in "a state of nature," where natural selection can result in the maintenance of human adaptations of benefit in local environments. Moreover, as you will recall from our discussion of human races, differences in body surfaces (which are obviously influenced by natural selection) are now largely irrelevant to our species because people adapt to the environment primarily by means of behavior, not by means of their body surfaces. But this does not mean that natural selection is no longer a major factor in human evolution. In fact, there are at least three major ways in which natural selection operates in human populations today—prenatal selection, postnatal selection, and fecundity selection. Let us now discuss briefly each of these ways in which natural selection is at work in human populations today and speculate as to how they may be affected by some existing and by some proposed eugenic measures.

Prenatal selection refers to any genetic and environmental factors that result in death sometime between fertilization and birth. As you know, prenatal death is frequently the result of genetic factors, such as abnormal chromosome constitutions or homozygosity for disadvantageous alleles; as such, it is beyond our present eugenic reach. Although it is true that genetic counseling and prenatal diagnosis must have some influence on prenatal selection in human populations, any eugenic effect so far has been, and will probably continue to be, meager.

Postnatal selection occurs when infants born alive fail to survive to reproductive age. In large part postnatal selection can be attributed to genetic factors. Included in this category are many of the serious inborn errors of metabolism and other genetic diseases that we have discussed before. Although the effects of modern medical treatment have been spectacular and of enormous benefit to some people and to their families, most geneticists agree that medical treatment has so far had little effect on the average gene frequencies in the global human population. Nonetheless, it must be admitted that the human species could be affected in a very adverse way by the "genetic load" of detrimental alleles that has already begun to build up because of human intervention and that will surely continue to do so in the future.

Fecundity selection, which may take place both within and between populations, occurs when some genetically distinct members of the population leave relatively more descendants than others. Fecundity selection *between* human populations at the present time favors people who live in the less highly developed regions of the world, especially in Latin America, Africa, and Asia, but this pattern may change before too long as effective means of birth control become more widely available. Fecundity selection *within* human populations also occurs, but is hard to measure and subject to frequent changes. Nonetheless, some elitist prophets of doom have made fecundity selection within the United States population a social issue. Their assertion is that the average intelligence of the U.S. population is decreasing because people of lower socioeconomic groups, who have, on the average, lower IQ scores, also tend to have the most offspring. Implicit in this argument is the contention that genetic, not environmental factors, are most important in determining how individuals perform on IQ tests, and, as we have discussed, that question is far from settled. Also, the biological significance, if any, to the human species of a decrease in IQ scores would be far from clear. Overall, fecundity selection, although it does occur both within and between human populations, provides us with little basis from which to predict the future course of human evolution.

Fecundity selection is also largely beyond the reach of existing eugenic measures, and for good reason. Eugenic measures applied to fecundity selection would necessarily require judgements about what constitutes "superior" genotypes and the selective breeding (or artificial insemination) of people who have traits judged to be desirable. Given the social climate of the day, such undertakings are not at all likely to materialize. (Recall that social pressures recently shut down the only XYY screening program in the U.S., in spite of a resounding vote of confidence from professors at the Harvard Medical School to continue the program.) One can only expect that genetic questions that have social overtones will continue to attract the interest of and to elicit reactions from various activist groups whose presence, for better or for worse, will surely strongly influence eugenic proposals of any kind.

In the end, the overall effect of natural selection within the human population today thus appears to be maintenance of the human species as it is at present. That is not surprising, because natural selection is usually a stabilizing influence that tends to put a brake on rapid or extreme evolutionary changes. This is because the extreme phenotypic variants within any population are much less likely to leave descendants than other members of the population.

Certain evidence in the fossil record also suggests a stabilizing influence of natural selection. The human species in its present form has been in existence for about 40,000 years, which though long by human standards is a mere flash in the pan of geological time. The fossil record shows that at least some trends in physical evolution, once they become established, continue for long periods of time, even as measured by geological standards. In general, they continue because they are directly or indirectly influenced by natural selection. One such trend among humans and other primates as well has been (within broad limits) the evolution of a more complicated and more capable brain, which is reflected in solid bone by an increase in cranial capacity—that is, by an increase in the volume of the brain. Yet in the last 40,000 years (a period of time in which somewhat more than 2,500 human generations have come and gone), the volume of the human brain has not changed at all. It is probably safe to assume that science fiction writers are misleading us when they conjure up images of our descendants hundreds of centuries hence, with overgrown brains encased in enormous globular skulls carefully balanced on pale, slender necks, or, worse yet, deposited in receptacles that fall far short of having the form and presence of a human body.

The fossil record also shows a slight tendency for an increase in height in the more recent stages of human evolution, but this is probably due mostly to environmental, rather than genetic, changes. Minor changes in human teeth, with a slight tendency for reduction in the number of wisdom teeth may also have occurred in the past 30,000 years or so, though this is less well documented than the change in stature.

Other than that, as judged by the fossil record, the physical evolution of human beings appears to be at a standstill, or at least proceeding at a rate that is so slow as to be unnoticeable. Of course, only time will tell, but most of us will probably be long dead before any noticeable physical changes in the human species occur. Nonetheless, we have all experienced social and other cultural changes, some of which occur so fast that they bewilder us. Whatever its final basis, whether genetic or environmental, whether measured by IQ scores or by height, whether judged by cranial capacity or by other measures, the fact is that human beings of all races are wonderfully variable. And this enormous stockpile of human differences is our species' greatest asset toward the future. *Variability*—whether physical, behavioral, biochemical, or otherwise—is what is important in the future evolution of any species, including our own. It is to be hoped and expected that people will always be at least as variable as they are today, and that by some means, be it natural or eugenic, people will continue to evolve, and will exist for a very long time to come.

Summary

Eugenics includes all programs and proposals whose aim is genetically improving the human species either by increasing the frequencies of desirable traits (positive eugenics), or by decreasing the frequencies of undesirable traits (negative eugenics). What is desirable or undesirable always depends on judgements of individuals, and differences of opinion concerning eugenic measures often lead to heated social controversy.

Genetic counseling and prenatal diagnosis have greatly benefited some individuals, but have had little effect on human genetics. Genetic engineering, perhaps by means of viruses containing "spliced-in" genes, will probably soon be a reality, but it is not likely to have any noticeable eugenic effects in the human population.

Heritability is a statistical concept that estimates how much of the phenotypic variation in any population is due to genetic factors (nature) as opposed to environmental factors (nurture). Although it is impossible to accurately assess the heritability of any human trait, the study of one-egg twins raised apart provides a good estimate of the relative importance of genetic and environmental factors.

Human behavior is so varied and so changeable that it is most often impossible to sort out the genetic and environmental factors that influence how a person behaves. Mental retardation, schizophrenia, and the ability to score high on IQ tests are behavioral traits known to be strongly influenced both by genetic and by environmental factors. Eugenic measures relating to human behavior engender the most heated controversies of all, as evidenced by the recent dispute over the significance of racial differences in IQ scores, and by the recent shutdown of a screening program for detecting XYY males.

Three types of natural selection are influencing the worldwide human population: prenatal selection, postnatal selection, and fecundity selection. The effects of natural selection on human body surfaces are no longer very important in human evolution, because people now adapt to the environment primarily by means of behavior, not body surfaces. Overall, natural selection acts as a stabilizing influence in human evolution, and our species has not undergone any noticeable physical change in the past 40,000 years at least. Yet social and other cultural changes in human behavior occur frequently and may take place very rapidly, as is well known to all of us. As long as people remain variable they will continue to evolve, and it is to be hoped and expected that they will do so for a very long time to come.

Suggested Readings

1. "Prenatal Diagnosis of Genetic Disease," by Theodore Friedmann. *Scientific American,* Nov. 1971, Offprint 1234. A discussion of amniocentesis and other diagnostic techniques that have both genetic and social aspects.
2. "The Manipulation of Genes," by Stanley N. Cohen. *Scientific American,* July 1975, Offprint 1324. Discusses some of the molecular details and some of the social implications of the newly devised technique of genetic engineering.
3. "Genetic Load," by Christopher Wills. *Scientific American,* March 1970, Offprint 1172. How the accumulated mutations of any species are usually detrimental but at the same time may be a priceless genetic resource.
4. "Intelligence and Race," by W. F. Bodmer and L. L. Cavalli-Sforza. *Scientific American,* October 1970, Offprint 1199. This recap of the differences in IQ scores between American blacks and whites concludes that the heritability of IQ scores cannot be accurately determined from the data presently available.
5. "Behavioral Implications of the Human XYY Genotype," by Ernest B. Hook. *Science,* Vol. 179, 12 Jan. 1973. A review of some of the data on this complicated subject.

APPENDIX

Some Genetics Problems Concerning Human Pedigrees

You may want to test your understanding of the patterns of inheritance discussed in Chapters 1 and 2 by working the following problems. The answers are at the end of the appendix.

1. Huntington's chorea is a rare, fatal disease of the nervous system whose symptoms are not exhibited until middle age. It is inherited as an autosomal dominant trait. Suppose that an apparently normal man in his early twenties learns that his father has just been diagnosed as having Huntington's chorea.
   a. What are the chances that the son will eventually develop the disease himself?
   b. If the son does not develop the disease, what are the chances that his offspring will have Huntington's chorea?
2. Two bald parents have four children, two bald and two with normal hair. Assuming that it is governed by a single pair of alleles, is this kind of baldness best explained as an example of dominant or of recessive inheritance?

3. Maple syrup urine disease is a rare inborn error of metabolism that derives its name from the odor of the urine of affected people. Those in whom the disease is untreated are severely mentally retarded, and they usually die as infants. Once it has appeared in a given family the disease tends to recur, but the parents of affected children are always normal. Assuming that a single pair of alleles governs the occurrence of this disease, what does this suggest about the inheritance of maple syrup urine disease?

4. About 7 percent of Caucasians in the United States cannot smell the odor of musk. If both parents cannot smell musk, all of their children will be unable to smell it. On the other hand, two parents who can smell musk generally have children who can also smell it; only a few of their offspring will be unable to smell musk. Assuming that a single pair of alleles governs this trait, what does this suggest about the inheritance of the ability to smell the odor of musk?[1]

5. Total color blindness is a rare condition that is inherited as an autosomal recessive trait. Affected people see the world only in shades of gray and can see best in dim light or in the dark. A woman whose father is totally color-blind intends to marry a man whose mother was totally color-blind. What are the chances that they will produce affected offspring?

6. As discussed in Chapter 1, albinism is inherited as an autosomal recessive trait. An albino man marries a normally pigmented woman and they have nine children, all normally pigmented. What are the genotypes of the parents and of the children?

7. A normally pigmented man whose father was an albino marries an albino woman whose parents were both normally pigmented. They have three children, two normally pigmented and one albino. List the genotypes of all these people.

8. A man who has sickle-cell trait marries a woman whose father had sickle-cell disease.
   a. Can they produce offspring who have neither sickle-cell disease nor sickle-cell trait?
   b. What proportion of their offspring will have sickle-cell disease?
   c. What proportion of their offspring will have sickle-cell trait?

9. A man of blood group O marries a woman of blood group A. The woman's father was of blood group O. What are the chances that their children will belong to blood group O?

10. In the following case of disputed paternity, which of the possible fathers can be excluded as the real father? The mother is of blood group B, the child is of blood group O, one possible father is of blood group A, and the other is of blood group AB.

11. Four babies were born in a hospital on a night in which an electrical blackout occurred. In the confusion that followed, their identification bracelets were mixed up. Conveniently, the babies are of four different blood groups:

---

1. From *Genetics, Evolution and Man,* by W. F. Bodmer and L. L. Cavalli-Sforza. W. H. Freeman and Company, 1976.

O, A, B, and AB. The four pairs of parents have the following blood groups: O and O, AB and O, A and B, B and B. Which baby belongs to which parents?[2]

12. A woman who has unusually short fingers marries a man who has fingers of normal length, and they have four children, two of each sex. One of their sons and one of their daughters have unusually short fingers.
    a. Draw the pedigree.
    b. If finger length is governed by a single pair of alleles, could the pedigree be explained by autosomal dominant inheritance?
    c. Could the pedigree be an example of autosomal recessive inheritance?
    d. Could the allele responsible for this trait be located on the X chromosome?

13. Assume that the following pedigree is for a trait that is rare in a particular population.

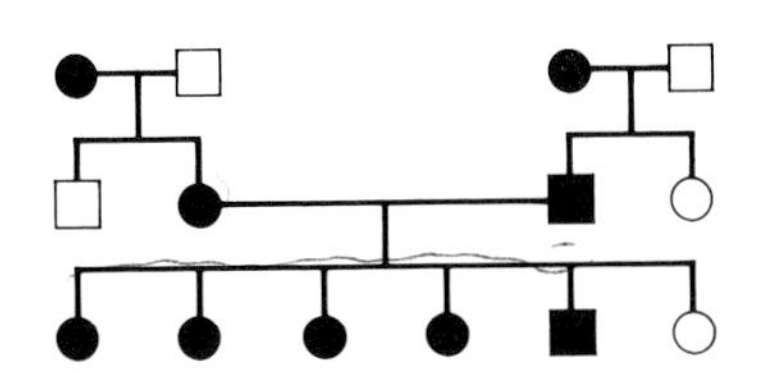

Indicate whether each of the following patterns of inheritance is consistent with or excluded by this pedigree.[3]
    a. Autosomal recessive.
    b. Autosomal dominant.
    c. X-linked recessive.
    d. X-linked dominant.
    e. Y-linkage.

14. Red–green color blindness is a relatively common condition that is inherited as an X-linked recessive. A normal woman whose father was red–green color-blind marries a man who has normal vision.
    a. What proportion of her sons would you expect to be red–green color-blind?
    b. If she married a man who was red–green color-blind, what proportion of their sons would you expect to have normal vision?
    c. If she married a man who was red–green color-blind, what proportion of their daughters would be carriers?

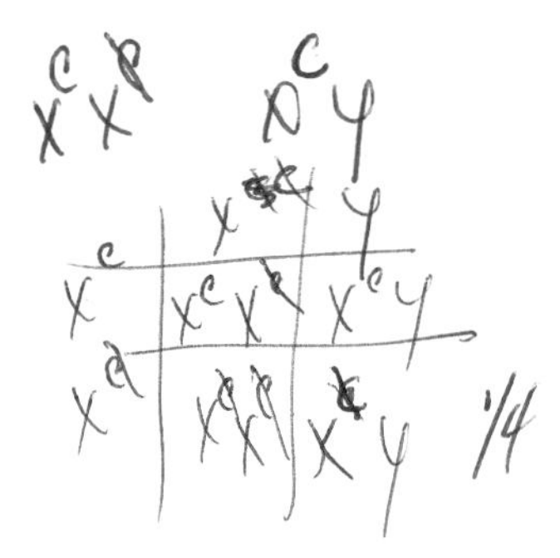

15. What pattern of inheritance *best* accounts for the following pedigree?

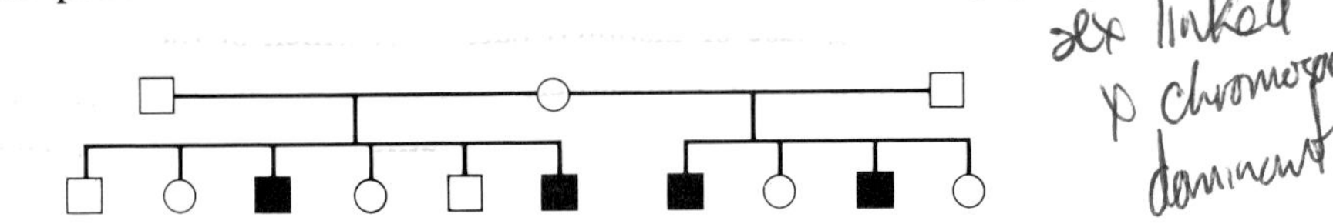

2. From *Heredity Evolution and Society*, by I. Michael Lerner and William Libby. W. H. Freeman and Company, 1976.

3. From *General Genetics*, by Adrian Srb, Ray D. Owen, and Robert Edgar. W. H. Freeman and Company, 1965.

16. As discussed in Chapter 2, hemophilia is inherited as an X-linked recessive trait. An apparently normal woman whose father was a hemophiliac marries a normal man.
    a. What proportion of their sons will have hemophilia?
    b. What proportion of their daughters will have hemophilia?
    c. What proportion of their daughters will be carriers?

17. Suggest three patterns of heredity that could account for the following pedigree. Which is least likely?

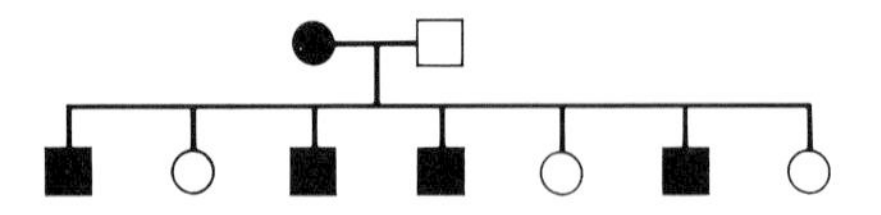

18. A man who is red–green color-blind (X-linked recessive) has four normal sons by his first wife and a color-blind daughter by his second wife. Both wives have normal vision. What is the genotype of each wife?

19. A short-fingered man (autosomal dominant) marries a woman who has normal hands and is red–green color-blind (X-linked recessive). List the possible phenotypes of their sons and daughters.

20. In the following pedigree a dot represents the presence of an extra finger and a shaded area represents the occurrence of an eye disease.[4]

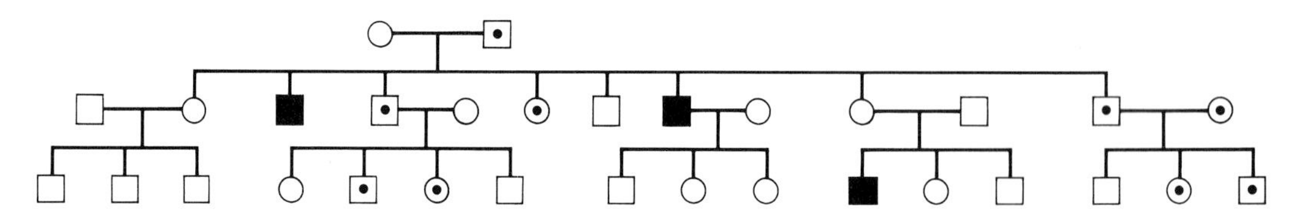

    a. What can you figure out about the inheritance of the extra finger?
    b. What two patterns of heredity might explain the inheritance of the eye disease?

21. Congenital deafness results when a person is homozygous for either or both of two recessive alleles, *d* and *e*. Both of the corresponding dominant alleles *D* and *E* are required for normal hearing, and the two pairs of alleles are inherited independently. A deaf man of genotype ddEE marries a woman with normal hearing who is of genotype *DdEe*. What proportion of their offspring will be deaf?

22. Complete the following table.

CONDITION	SEX CHROMOSOMES	NUMBER OF BARR BODIES
Normal woman		
Normal man		
	XO	
	XXY	
Down's syndrome		

4. From *An Introduction to Genetic Analysis,* by David T. Suzuki and Anthony J. F. Griffiths, W. H. Freeman and Company, 1976.

23. Would you expect a person of sex chromosome constitution XXXX to have higher, lower, or approximately equal concentrations of G6PD as compared to a person whose sex chromosomes are XXXY?

24. A woman who has Turner's syndrome is found to have hemophilia, yet neither of her parents have the disease. How is this possible?

25. How many Barr bodies would be found in a person who is a sexual mosaic with sex chromosomes XX/XXXY?

Answers

1. a. The chances are 50:50. Because the disease is rare, the affected father is most likely heterozygous (of genotype *Hh*, *H* for Huntington's chorea). All of his offspring therefore have a 50:50 chance of inheriting the dominant allele.
   b. Virtually zero, assuming that he marries an unaffected woman. Huntington's chorea can arise by spontaneous mutation, but this is a very rare event.

2. Dominant. If some of the children are not bald, the parents must be heterozygous for the trait (of genotype *Bb*.) If this kind of baldness were recessive, both parents would have to be homozygous recessive (of genotype *bb*); thus they could produce bald offspring only.

3. This pattern suggests that maple syrup urine disease is inherited as a recessive trait.

4. This pattern suggests that the ability to smell the odor of musk is dominant and that the inability to do so is recessive.

5. The chances are one in four. The man and woman must both be heterozygous (of genotype *Cc*,) so these are the kinds of offspring they can produce:

	C	*c*
C	*CC*	*Cc*
c	*Cc*	*cc*

   Thus, on the average, one-fourth of their offspring would be normal (*CC*), one-half would be carriers (*Cc*), and one-fourth would be totally colorblind (*cc*).

6. The man is an albino, so he must be homozygous recessive (of genotype *aa*). None of the nine children is an albino, but each must be a carrier (of genotype *Aa*). Since there are no albinos among the nine children, the mother is apparently of genotype *AA*.

7. The man must be of genotype *Aa*, and his father of genotype *aa*. Similarly, the woman must be *aa* and her parents must both be *Aa*. Two of the children are of genotype *Aa*, and the other is *aa*.

8. a. Yes. Both parents must be heterozygous, so they can produce the following kinds of offspring:

	Hb^S	Hb^A
Hb^S	Hb^SHb^S	Hb^SHb^A
Hb^A	Hb^SHb^A	Hb^AHb^A

(Hb^AHb^A = normal, Hb^SHb^A = sickle-cell trait, Hb^SHb^S = sickle-cell disease.)

b. one-fourth (genotype Hb^SHb^S).

c. one-half (genotype Hb^SHb^A).

9. The chances are 50:50. The woman must be of genotype I^AI^O, so these parents can produce the following offspring:

	I^A	I^O
I^O	I^AI^O	I^OI^O
I^O	I^AI^O	I^OI^O

Thus about one-half of their offspring will be of genotype I^OI^O and will therefore belong to blood group O.

10. The man of blood group AB cannot be the real father. The mother is apparently of genotype I^BI^O. Her child is of group O and therefore must have inherited the I^O allele from each parent. The man of group AB can contribute only I^A or I^B and therefore cannot be the father. The man of group A could be of genotype I^AI^O, but he could also be of genotype I^AI^A. Thus we can neither exclude the man of group A nor conclude that he is the real father.

11. a. Baby O could only belong to parents O and O, because the parents must be of genotype I^OI^O.

b. Baby AB must belong to parents A and B because only they could produce an offspring of genotype I^AI^B.

c. Of the remaining two, baby A cannot belong to parents B and B, because they must be of genotype I^BI^B ro I^BI^O. Baby A must therefore belong to parents AB and O.

d. Hence, baby B must belong to parents B and B.

12. a. The pedigree is this:

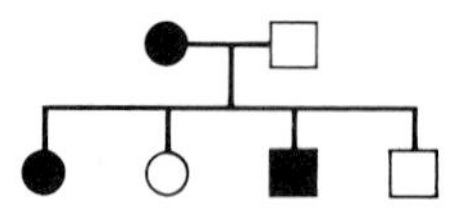

b. Yes. In fact, this trait (officially known as *brachydactyly*) was the first autosomal dominant trait described among human families.

c. Yes. But this is not likely to be an example of autosomal recessive inheritance because the father would have to be a carrier, which is unlikely.

d. Yes. The trait cannot be inherited as an X-linked recessive because one of the sons does not manifest the trait. But the pedigree is consistent with X-linked dominant inheritance.

13. a. Autosomal recessive inheritance is excluded because the mating of two affected people produced a normal daughter.
    b. The pedigree is consistent with autosomal dominant inheritance.
    c. X-linked recessive inheritance is excluded because an affected woman produced a normal son.
    d. X-linked dominant inheritance is also excluded because an affected man produced a normal daughter.
    e. Y-linkage is excluded because of the presence of affected women.

14. a. One-half. The woman must be a carrier. All sons receive one or the other of their mother's X chromosomes, so the chances are 50:50 that any son would receive the chromosome with the abnormal allele.
    b. One-half. No sons receive their father's X chromosome, so the fact that the father is affected does not alter the chances that his sons will be red–green color-blind.
    c. One-half. All daughters receive thier father's X chromosome and one or the other of their mother's X chromosomes. Thus this marriage would produce color-blind and carrier daughters in roughly equal proportions.

15. X-linked recessive inheritance. The pedigree could also be explained by autosomal recessive inheritance, but that would require all three parents to be carriers of the same allele.

16. a. One half. The woman must be a carrier of hemophilia, and on the average half of her sons get the X chromosome that has the abnormal allele.
    b. None of them. An affected daughter could be produced only if the father were affected.
    c. One-half. On the average half of the daughters receive the abnormal X chromosome from their mother and therefore are carriers.

17. The pedigree is probably an example of X-linked inheritance, either dominant or recessive. Autosomal dominant inheritance cannot be excluded, but is least likely because all sons and no daughters are affected.

18. The first wife is most likely normal, and the second wife must be a carrier of red–green color-blindness.

19. All sons will be red–green color-blind and about half of them will have short fingers. All daughters will be carriers of red–green color blindness and about half of them will have short fingers.

20. a. The extra finger is probably an example of autosomal dominant inheritance.
    b. The eye disease is probably inherited as an X-linked recessive trait, but it could be an example of autosomal recessive inheritance.

21. One-half. All of the man's sperm cells must contain the factors *d* and *E*. But the woman can produce four different kinds of eggs, namely: *DE*, *De*, *dE*, and *de*. So we can construct the following table:

	dE	
DE	*DdEE*	(normal)
De	*DdEe*	(normal)
dE	*ddEE*	(deaf)
de	*ddEe*	(deaf)

22. The completed table:

CONDITION	SEX CHROMOSOMES	NUMBER OF BARR BODIES
Normal woman	XX	1
Normal man	XY	0
Turner's syndrome	XO	0
Klinefelter's syndrome	XXY	1
Down's syndrome	XX or XY	1 or 0

23. The concentrations of G6PD in these two persons would be approximately equal. The XXXX individual has one active X chromosome and three Barr bodies, and the XXXY individual has one active X chromosome and two Barr bodies.

24. This could happen if the woman's mother were a carrier and nondisjunction resulted in the daughter's receiving no X chromosome from her father. If the daughter received only her mother's abnormal X chromosome, she would show symptoms of hemophilia.

25. One or two Barr bodies would be found, depending on which cell line is examined. Some of this person's cells are XX and therefore have 1 Barr body, whereas other cells are XXXY and therefore have 2 Barr bodies.

Index

References to illustrations are given in italic type.